NUCLEAR FACILITY THREAT ANALYSIS AND TACTICAL RESPONSE PROCEDURES

NUCLEAR FACILITY THREAT ANALYSIS AND TACTICAL RESPONSE PROCEDURES

By

JERRY J. CADWELL, J.D.

Department of Nuclear Energy
Brookhaven National Laboratory
Upton, New York

CHARLES C THOMAS • PUBLISHER

Springfield • Illinois • U.S.A.

Published and Distributed Throughout the World by

CHARLES C THOMAS • PUBLISHER
2600 South First Street
Springfield, Illinois, 62717, U.S.A.

ISBN 0-398-04778-2

Library of Congress Catalog Card Number: 82-16801

With THOMAS BOOKS *careful attention is given to all details of manufacturing and design. It is the Publisher's desire to present books that are satisfactory as to their physical qualities and artistic possibilities and appropriate for their particular use.* THOMAS BOOKS *will be true to those laws of quality that assure a good name and good will.*

Printed in the United States of America

I-R-1

Library of Congress Cataloging in Publication Data

Cadwell, Jerry J.
Nuclear facility threat analysis and tactical response procedures.

Bibliography: p.
1. Nuclear facilities--Security measures. I. Title.
TK9153.C3 1983 621.48'35'068 82-16801
ISBN 0-398-04778-2

PREFACE

THIS book is intended to aid the management of a nuclear facility security force in making plans for, and training guards to make, practical alarm assessments, and to provide model response procedures and information that can be used to implement security plans as required in Part 10 of the U.S. Code of Federal Regulations and Part 73.55 Appendix C. The author found that the available literature does not offer sufficient detail in the areas of threat analysis and tactical response. It is the author's hope that this book will provide some help to the nuclear facilities in making up their security contingency plans.

Substantial comments and assistance were provided by Dr. John H. Cusack, Head of the Brookhaven National Laboratory Technical Support Organization and Dr. John N. O'Brien of Brookhaven National Laboratory. A number of individuals, too numerous to identify, at several nuclear facilities were kind enough to provide source material that was used in this book. Generous suggestions and guidance were received from Dr. Joseph P. Indusi and Ms. Julia M. Smith, both of Brookhaven National Laboratory. All pictures were provided by Allied-General Nuclear Services at Barnwell, North Carolina, courtesy of Mr. E.L. Musselwhite and his superb organization of Security Inspectors.

This book is a compilation of reports originally submitted to the United States Nuclear Regulatory Commission by the author and others of Brookhaven National Laboratory, Upton New York, which are now publicly available in the United States

Nuclear Regulatory Commission Public Document Room, Washington, D.C.

Jerry J. Cadwell, J.D.
Shoreham, L.I., New York

INTRODUCTION

OVERVIEW

THE security recommendations which follow primarily address the potential threat posed by terrorist actions that may lead to the sabotage of a nuclear facility such as a nuclear electric power station or a U.S. Department of Energy nuclear facility. Sabotage of a nuclear facility has the potential to cause the release of radioactive materials into the atmosphere with consequent property damage and loss of life. Such terrorist actions have been characterized by the U.S. Nuclear Regulatory Commission in the U.S. Federal Code of Regulations:

> (i) A determined violent external assault or attack by stealth or deceptive actions, of several persons with the following attributes, assistance and equipment: (A) well-trained (including military training and skills) and dedicated individuals, (B) inside assistance which may include a knowledgeable individual who attempts to participate in both a passive role (e.g., provide information), an active role (e.g., facilitate entrance and exit, disable alarms and communications, participate in violent attack), or both (C) suitable weapons, up to and including hand-held automatic weapons, equipped with silencers and having long range accuracy, (D) hand carried equipment, including incapacitating agents and explosives for use as tools of entry or for otherwise destroying reactor, facility, transporter, or container integrity or features of the safeguards system, and
>
> (ii) An internal threat of an insider including an employee (in any position).[1]

OBJECTIVE

The objective of this book is to provide recommendations in the form of procedures describing how a nuclear guard force should initially react* to alarms, assess alarms, determine threat level, and make a tactical response to threats.

Certain alarms require an immediate tactical response while a threat determination is being simultaneously made; however, most alarms require a tactical response to be made only after the alarm has been assessed and determined to be a threat.

This book develops procedures that describe the initial reaction activity required to place a guard into a position where he or she may evaluate an alarm. Procedures for determination of the threat level are developed to deal with the question, "After having gotten into position to evaluate the alarm, how does one determine the threat level represented by the alarm?" This approach of guard response† and threat analysis is described schematically in Figure 1. Procedures required to describe the mobilization and tactical response to an actual threat and presented in this book are designed to place a guard or guards in a position to intercept, delay entry, or arrest an intruder. In addition, these procedures are designed to place a guard or guards in position to confront an intruder within 1 to 3 minutes after the alarm, while maintaining security throughout the plant.[2, 3] These generic procedures are developed with the objective of providing as much help as possible to security experts in developing their own specific procedures for their own security plan.

It is assumed that a threat or attempt to commit radiological sabotage at a nuclear facility may occur at any time with or without prior warning. It is also assumed that an adversary action can be detected by proper surveillance of alarms, barriers, protected areas, and vital areas. During a nonroutine safeguards contingency it is assumed that an adversary's primary intent is to commit radiological sabotage.

The procedures provided in this book are intended to provide guidance for guard force response once an adversary has been detected. Techniques specifically designed to deal with the threat posed by a knowledgeable insider are not discussed.

---

*An initial reaction to an alarm is that activity required to place a member of the reactor guard force into a position to evaluate (assess) the meaning of an alarm.

†Unless otherwise noted, "response" means "tactical response" throughout this book.

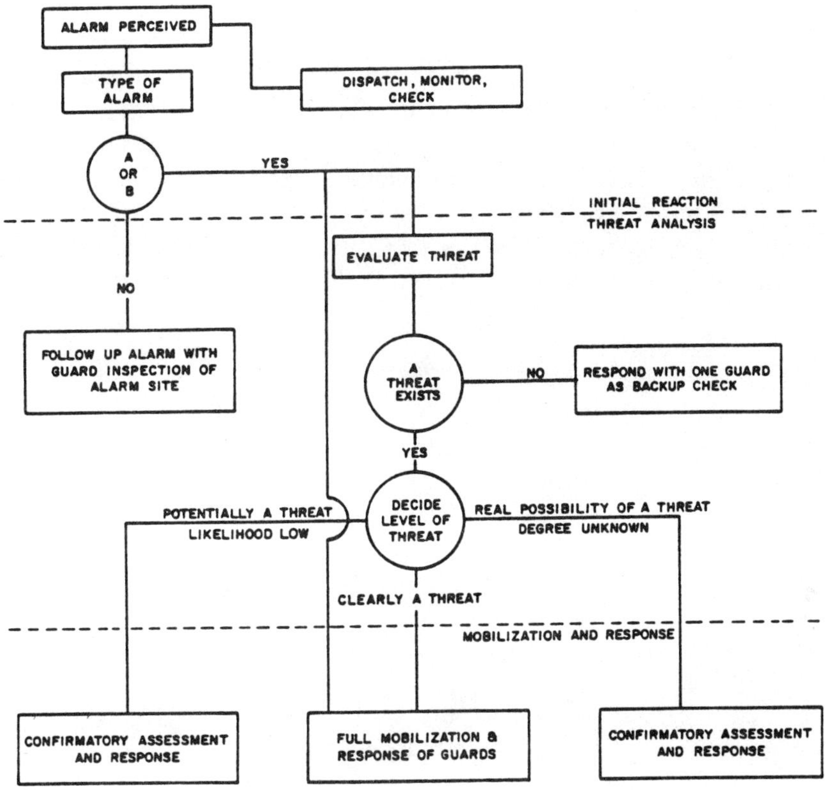

Figure 1. Flow Diagram for Threat Analysis and Tactical Response

ALARMS CONSIDERED

Central to the concepts developed in this book is a list of alarms. This list was developed to cover all the concerns expected to be encountered as a result of terrorists attempting radiological sabotage or attempted theft of special nuclear material (SNM).

The titles of the alarms are descriptive and are meant to describe the activity of the intruder. Strictly speaking, an "alarm" is a buzzer sounding or a bell ringing and not the activity itself. However, to simplify the approach taken in this book, an alarm is considered as notice to the recipient that the activity represented by the title of the alarm is taking place or probably is taking place. For instance, a fence motion detector alarm signals that a fence is being moved. It does not tell what, who, or why the fence is being moved.

Most alarms are either visual or audio. Alarms given as a result of visual sighting are those which, when noticed, give a great deal of information at the instant of observation. This distinction becomes important when it is recognized that a visual alarm is a notice that some unauthorized activity probably is underway. An audio alarm is notice that an unauthorized activity may or may not be underway.

CONTENTS

NUCLEAR FACILITY THREAT ANALYSIS AND TACTICAL RESPONSE PROCEDURES

Chapter 1

INITIAL REACTION TO AND THREAT ANALYSIS OF ALARMS

NOT all alarms are equal in terms of the possible security contingency they announce. As an aid in developing a determination of threat level, each alarm is classified by the author according to the relative danger it presents. This hierarchy of alarms was derived by considering the subjective relative threat of each alarm. The author's knowledge and field experience in the areas of reactor operation, nuclear safeguards, and police science constituted the basis for the subjective opinions.

The philosophy adopted by the author of this book is that the response of the guards should be tailored to the threat level indicated by the alarm. Therefore the alarms are divided into four classes; each alarm announces one of the following: (A) clearly a threat of substantial magnitude, (B) clearly a threat, (C) a real possibility of a threat, degree unknown, or (D) potentially a threat, likelihood low.

Each alarm should be viewed as a threat until proven otherwise. For Class A and B alarms, which clearly represent a threat, no initial reaction or threat analysis (alarm assessment and threat level determination) need be made before a tactical response. The clear-threat-alarm requires a preplanned tactical response. During and after this preplanned response, there would be an alarm assessment to ascertain what further response was required and to determine if the threat could materialize into a confrontation or sabotage.

The Class C and D alarms should be regarded as something less than a maximum threat until there is reason to believe that the guard response should be escalated to higher levels. Based on available information,[4] experience has been that most or nearly all Class C and D alarms are not an indication or a real threat. However, the potential for damage is so high that there should be something more than a routine response.

ALARM CLASSIFICATION

The author considers the following as Class A and Class B alarms. These alarms are regarded as threats requiring immediate guard mobilization and tactical response in conjunction with the threat analysis. They give a warning that a situation exists which could result in the sabotage of a nuclear facility of sufficient magnitude to be a danger to the health and safety of the public.[5]

Class A Alarms

1. Someone in vital area—no apparent authorization.
2. Vehicle forcing its way into protected/vital area.
3. Vehicle forcing its way out of protected/vital area.
4. Hardware/explosives viewed (sabotage aids).

Class B Alarms

5. Someone coming over the fence.*
6. Someone going over fence away from plant.
7. Component damage detected (alarm or operations).
8. Sound of explosion.
9. Door alarm (certain doors).
10. Personnel detector alarm.
11. Floor weight detector alarm (certain places).
12. Guard-requests-aid alarm.
13. Area radiation alarm.

Most of the alarms listed above are indicators of actual intrusions; however, some of the alarms only infer intrusions with a high degree of confidence. These alarms may be the last indication of the physical presence of personnel in vital areas and may represent an intruder who has penetrated the security system to the

---

*Throughout this book unless otherwise denoted, "fence" means the protected area perimeter fence.

point that sabotage or theft of SNM is imminent. Thus, inferred intrusions must be treated as actual intrusions. Alarms listed above which infer intrusions are as follows:

9. Door alarm.
10. Personnel detector alarm.
11. Floor weight detector alarm.
12. Guard-requests-aid alarm.
13. Area radiation alarm.

Class C and D alarms generally require guard mobilization and tactical response only after an alarm assessment and determination of threat level have been performed.

Class C Alarms

14. Someone coming toward or near fence.
15. Someone in protected area – no apparent authorization.
16. Someone going away from fence.
17. Suspicious demeanor viewed.
18. A barrier breached – hole in fence/door open.
19. Omni-detector alarm.
20. Water intake detector alarm.
21. CCTV motion detector alarm.
22. Audio listening devices alarm.
23. Metal detector/explosive detector alarm.
24. Bomb/extortion threat.
25. Guard strike.
26. Kidnap – threatened or actual.
27. Security radio frequency jammed.
28. Civil disturbance.

Class D Alarms

29. Fence motion detector alarm.
30. Radiation alarm at portal.
31. Fire alarm.
32. Life-threatening-accident alarm.

A review of the number of alarms versus class of threat (Figure 2) shows that 28 out of 32 alarms are considered as clearly a threat (Class A or B) or at least a real possibility of a threat (Class C). Based on this classification of the possible threat level represented by each alarm, a flow diagram (Figure 3) was constructed to link the timeliness of an assessment of an alarm to the level of threat the alarm represented. Figure 3 provides a way of thinking which would aid in developing the speed of response for each threat represented by an alarm.

Hierarchy Class	Total Alarms Per Class
A. Clearly a Threat of Substantial Magnitude	4
B. Clearly a Threat	9
C. Real Possibility of a Threat — Degree Unknown	15
D. Potentially a Threat — Likelihood Low	4
TOTAL	32

Figure 2.

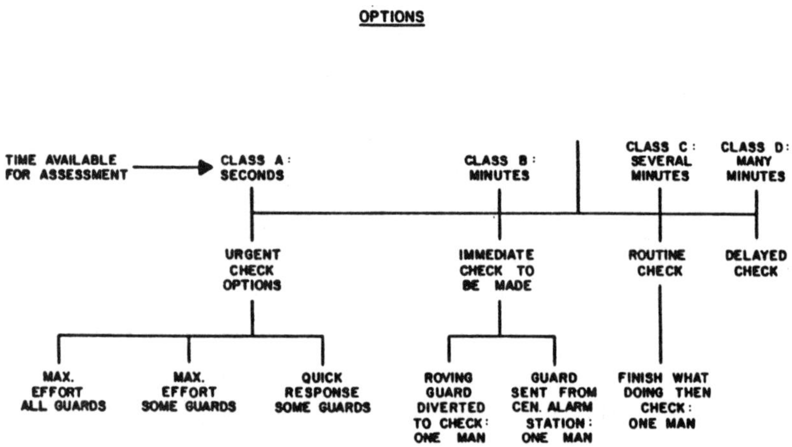

Figure 3. Assessment of Alarm

RECOMMENDATIONS FOR INITIAL REACTIONS TO ALARMS

Assessment of alarms and determination of threat level are closely associated with the initial reaction to alarms. For Class A and Class B alarms, *clear threats,* the threat analysis is concurrent with the initial reaction and tactical response. Class C and Class D alarms require a tactical response only after the initial reaction and

Figure 4. Checking Fence Alarms to be Sure They Are Functioning. Courtesy of E.L. Musselwhite and Allied-General Nuclear Services, Barnwell, North Carolina.

alarm assessment has determined that a threat exists. The following techniques of initially reacting to alarms are recommended:

Class A and B Alarms

- Initiate guard mobilization and tactical response concurrent with the initial reaction and threat analysis. (Alarms 1-13)

Class C and D Alarms

- Send a guard to inspect and assess the alarm. (Alarms 14, 15, 17, 22, 23, 28, 30-32)

- Send a guard to the protected area side of the fence; guards in a vehicle go to the exterior side of the fence. (Alarm 16)
- Deploy guards to the protected area to observe. (Alarm 18)
- View area with CCTV, if possible, then send a guard. (Alarms 19-21, 29)
- Respond to management request for services. (Alarms 24-26)
- Call the Central Alarm Station (CAS). (Alarm 27)

RECOMMENDATIONS FOR THREAT ANALYSIS OF ALARMS

For this analysis, "determination of the level of a threat announced by an alarm" is defined as, having got into position to

Figure 5. Using the Computer. Courtesy of Allied-General Nuclear Services.

evaluate the alarm (initial reaction to alarm completed), what does the guard or guard force supervisor do to determine the level of threat? The following general recommendations are made regarding determination of threat level.

The security supervisor makes a determination of threat level based on verbal information obtained from an on-the-scene viewer and information obtained at the central alarm station. The data needed by the security supervisor would include the following:

1. number of alarms received and history of recent alarms (source of information: Central Alarm Station [CAS]),
2. location and nature of alarm(s) received (source: CAS),
3. areas covered by closed circuit television (CCTV) surveillance (source: physical protection plan),
4. whether or not intruder* is sighted (source: CCTV or person investigating cause of alarm),
5. visual evidence of intrusion (other than sighting of intruder),
6. cause for a false alarm (source: varied),
7. test and inspection report on intrusion alarm equipment (source: instrument mechanic),
8. visual assessment of strength of intrusion (source: CCTV or person investigating cause of alarm),
   (a) number of people involved,
   (b) ages (adult/child), sex, and physical characteristics,
   (c) apparent intent or aggressiveness of entry (is it just a misunderstanding or prank, or really an armed act of aggression),
   (d) demeanor,
   (e) arms, equipment, and tools,
   (f) location of intruder (position relative to vital areas or vital equipment, or approximate distance to licensee's property line),
   (g) movement of individual (speed and apparent destination on-site or direction of travel),
   (h) visible badge identification,

*Unless otherwise denoted, "intruder" and similar terms can mean one or more than one individual throughout this book.

(i) vehicles,
(j) degree of intoxication,
(k) reaction to verbal contact and request to halt,
(l) likelihood of guard's ability to restrain individual without resort to use of deadly force,
(m) activity in which engaged (fishing, setting explosive, etc.),
(n) likelihood of malevolent acts having been committed by exiting individual.

The logic of alarm assessment[6] of alarms perceived at the central alarm station is shown in Figure 6. The *decision criteria* in Figure 6 are as follows:

1. Is there any immediate confirmation of danger?
   - "YES" if redundant alarm signals have been received at central alarm station.
   - "YES" if intrusion activities can be seen on CCTV monitor.
   - "YES" if a guard report is received at same time as or just after alarm signal.
   - "YES" if noise of gunfire or explosion is heard in central alarm station.
   - "NO" if none of the above.
2. Is an intrusion imminent or in progress?
   - "YES" if person investigating cause of alarm sees intruder.
   - "YES" if person investigating cause of alarm sees evidence of intrusion such as hole in fence, guard shot, strange vehicle within protected area, etc.
   - "NO" if none of the above.
3. Was the alarm sensor triggered by an artifact?
   - "YES" if artifact can be seen (stray dog, groundhog, etc.) or demonstrated by testing (wind, snow, etc.).
   - "NO" if none of the above.
4. Is the equipment malfunctioning?
   - "YES" if set point or calibration is found to be in error.
   - "YES" if equipment test shows equipment to be faulty.
   - "NO" if none of the above.

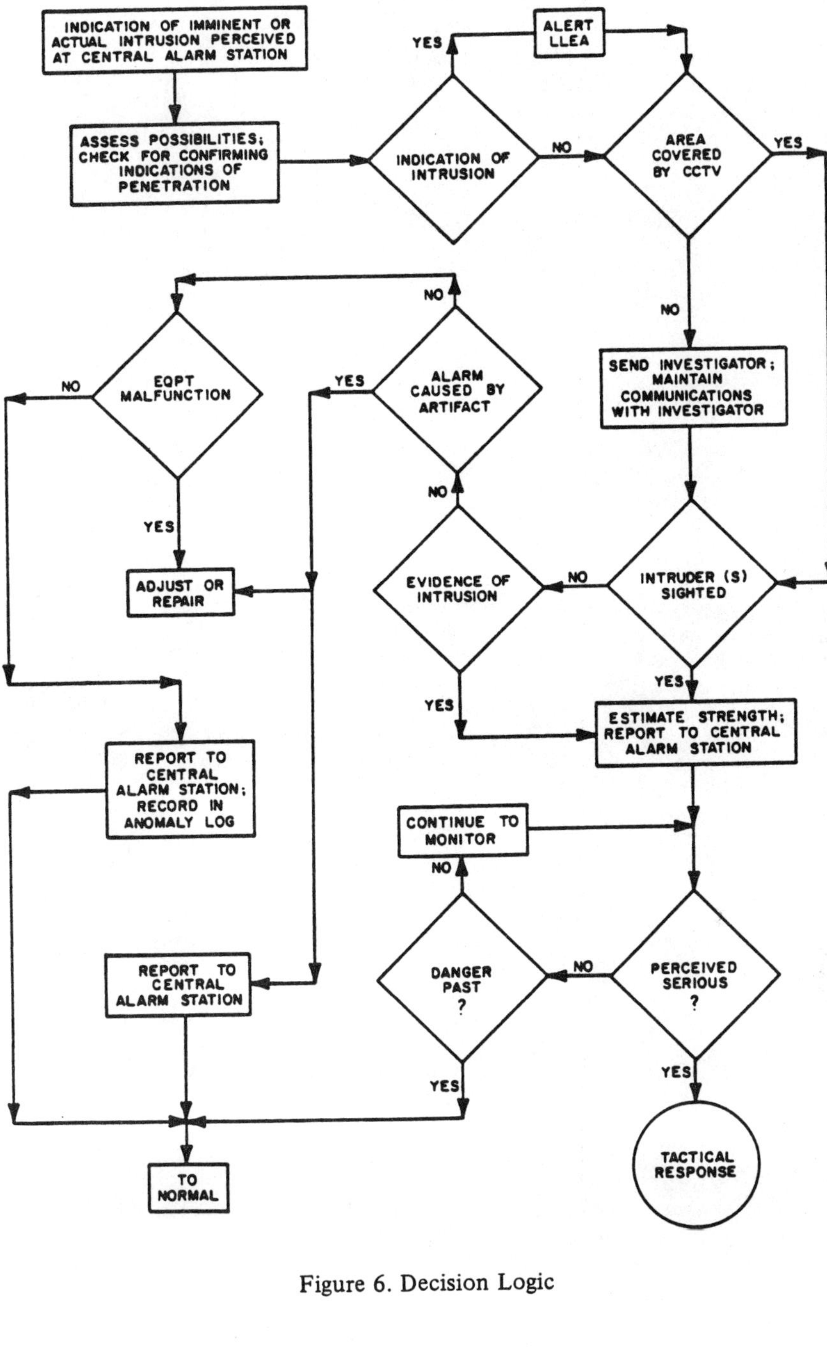

Figure 6. Decision Logic

The *action criteria* required to support the logic of Figure 6 for determination of threat assessment are as follows:

1. Assess possible significance of perceived indication.
   If area covered by alarm sensor has CCTV surveillance, check CCTV monitor. If area can be seen by a guard, have guard report on what (if anything) he can see.
2. Maintain communications with person(s) investigating cause of alarm.
   Until assurance is received that danger does not exist, person(s) investigating should be required to maintain continuous radio contact or call in every ten seconds; loss of contact should be taken as evidence that investigator has met foul

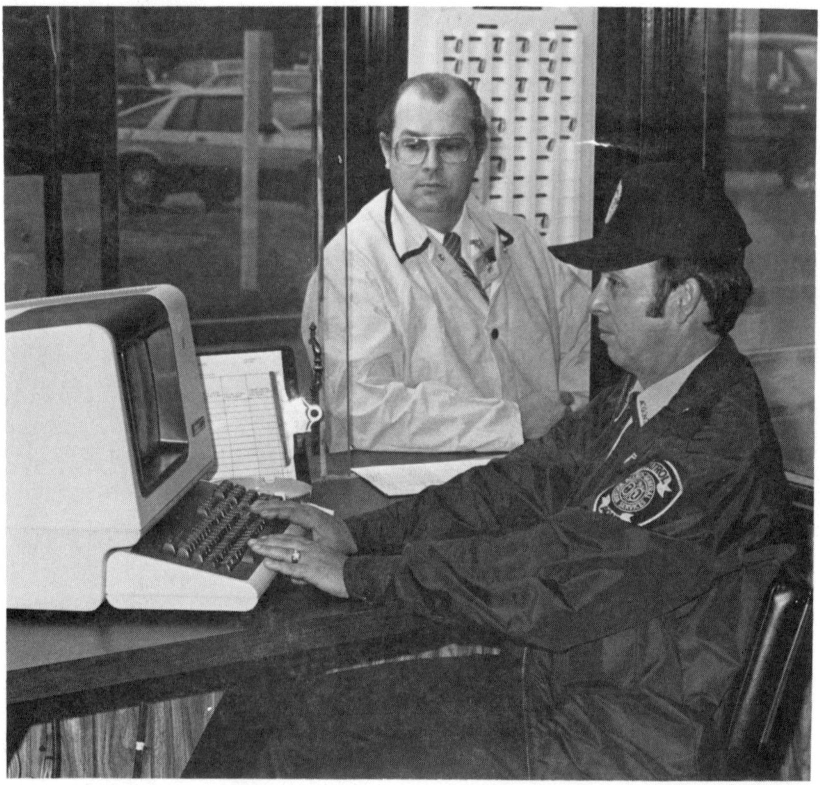

Figure 7. Using the Computer to Check Entrance Request. Courtesy of Allied-General Nuclear Services.

play from the intruder.

3. Confirm or nullify validity of alarm using the *decision criteria* given above.
   Investigator should first check for possibility of intrusion in progress. If no evidence of intrusion can be found, investigator should then check for triggering of sensor by an artifact. If neither of the above apply, sensor and associated circuitry should be checked for equipment malfunction. If none of the above are verified, the incident should be logged in the record of anomalies.
4. Adjust or repair malfunctioning equipment.
   Use standard calibration, test, maintenance, and repair procedures for equipment involved. If equipment will be out of service for more than a half-hour provide compensatory protection (additional guards, watchmen, backup equipment, etc., as appropriate).
5. Notify the local law enforcement agency (LLEA) before investigation if confirmatory indications are received.
6. Determine extent to which intrusion has taken place.
   If investigator sees evidence of intrusion, he should report the probability of the number of intruders in certain areas:
   (a) those still outside of outer barrier,
   (b) those within protected area,
   (c) those probably within protected area or vital area (if intruders are not seen but evidence of their presence is noted).
7. If intrusion has occurred, estimate strength of intrusion.
   Investigator should note these facts:
   (a) number of persons apparent,
   (b) whether adults or children,
   (c) whether armed (if possible to determine),
   (d) state of sobriety (if possible to determine),
   (e) any other data pertinent to perception of magnitude of danger.

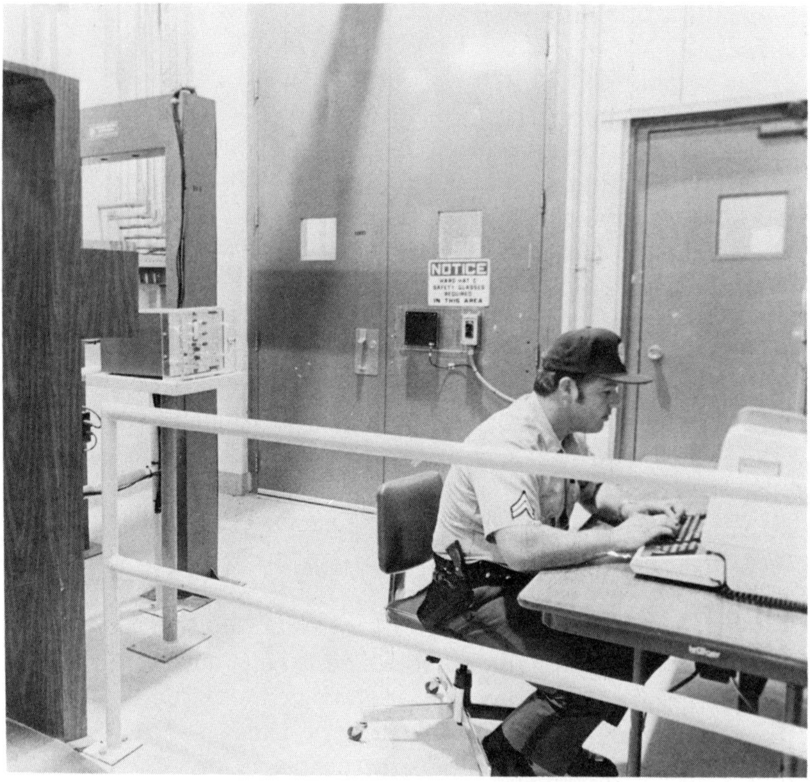

Figure 8. Monitoring Alarms Outside an Entrance with a Metal Detector Doorway. Courtesy of Allied-General Nuclear Services.

Chapter 2

TACTICAL RESPONSE TO ALARMS AND THREATS

BEFORE dealing with the specific tactical response required for each alarm, the various general types of response techniques available for use must be developed and considered.

Since the primary objective of the tactial response is to protect the nuclear facility from sabotage or theft of special nuclear material while minimizing the possibility of the intruder taking a hostage and to a lesser degree to protect those people in the plant, the guard personnel should clearly choose to protect the plant over engaging the intruder in a combat situation if the choice is presented.

The greatest potential loss that could be caused by intruder is sabotage or theft of special nuclear material affecting the health and safety of the public. Losses of lesser magnitude include loss of life by a guard, employee or intruder. There are less direct losses that may occur to the licensee long after the incident: liability suits and legal costs; imprisonment of guard as a result of improper actions by the guard; possible loss of life as a result of a successful murder or manslaughter charge; and consequent bad publicity to the local utility and the the nuclear industry.

In the event of an intrusion the nuclear facility management must make a choice of actions while considering the likelihood of an attack, the likelihood of a successful attack, the consequences of a successful attack, and the potential immediate and after-the-fact losses.

To establish a guard response technique one must consider

several things:

1. how many guards are available to respond at the time needed,
2. how fast can these guards respond,
3. where are the guards needed,
4. what do these guards do when they arrive where needed?

Certain activities of the plant require guards to remain in place even if other parts of the plant are under attack. For instance the plant cannot be left unguarded in one region because all guards are responding to a tactical response in another section of the plant. Figure 10 is an attempt to bracket the number of guards who could react to an order to respond. The data obtained from Figure 10 is conjectural and only provides some outside limits to the number of guards one could expect to respond.

Figure 11 gives a schematic of where guards typically are located in a nuclear power plant. These locations show the widely varying distances and places from which the guards will have to respond. The obvious result is that the guards will arrive at different times making a "quick" concerted response difficult at best.

After the central alarm station has decided to send a guard to investigate an alarm, the basic sequence of confrontation begins by sending the guard to view the situation as shown in Figure 12. With the aid of additional guards, the guard confronts the intruder as shown in Figure 13 and takes the intruder into custody and conducts a search as shown in Figure 14.

General tactical responses to an intrusion are treated in the discussions and schematics below.

GUARD TACTICAL RESPONSE IN GENERAL

The following seven tactical responses illustrate some of the generalized approaches that can be modified or used in combination as the situation demands. For example, the response of a "single confrontation with a reserve" could be revised to include two guards in a vehicle who confront a trespasser who is on licensee property but is outside of the protected area fence.

Figure 9. Facility Bomb Squad Experts Doing an Investigation. Courtesy of Allied-General Nuclear Services.

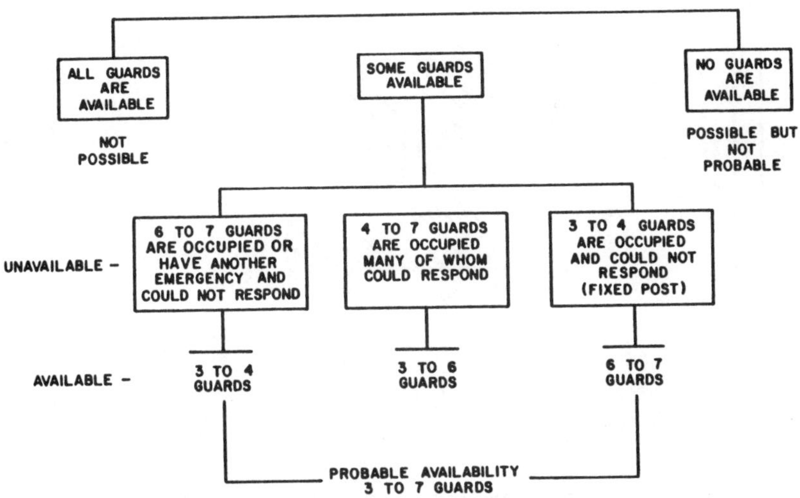

Figure 10. Number of Guards Available to Respond

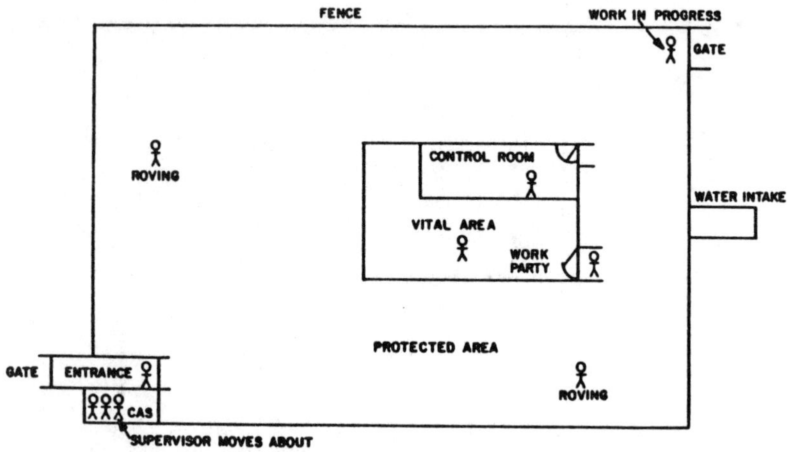

Figure 11. Location of Ten Armed Guards in Nuclear Facility

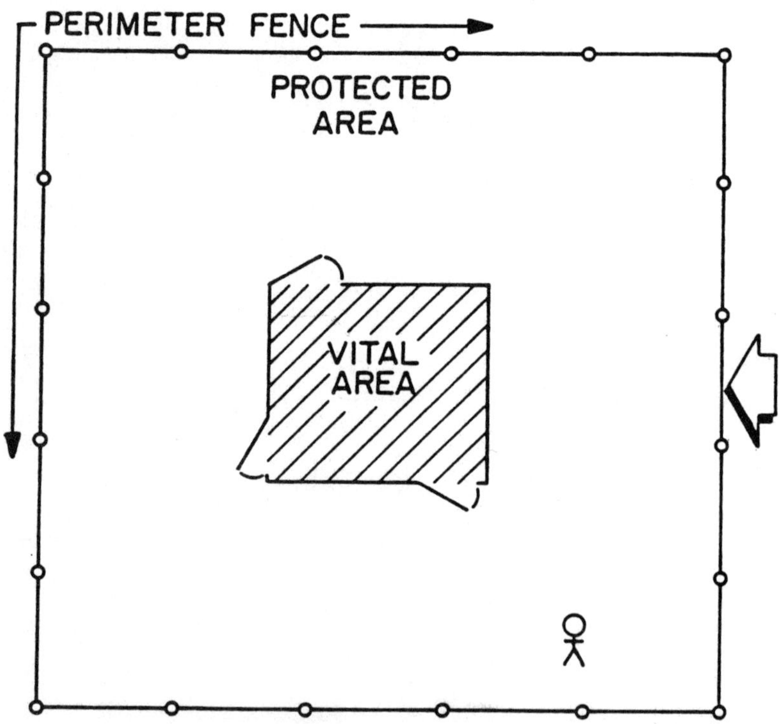

Figure 12. Single Guard and Intruder

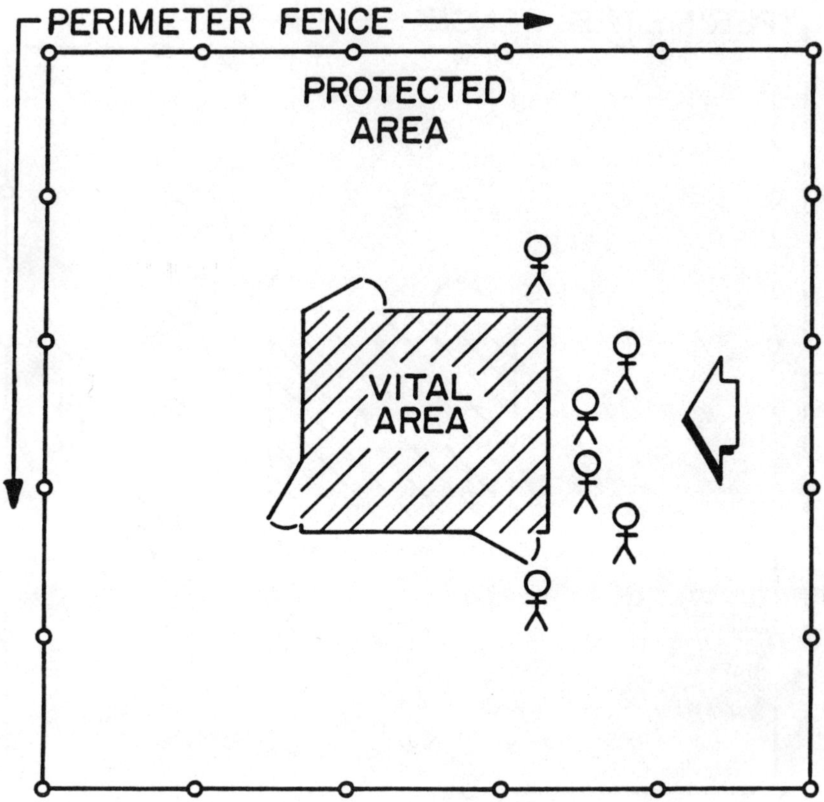

Figure 13. Guard Force in Show of Superior Strength

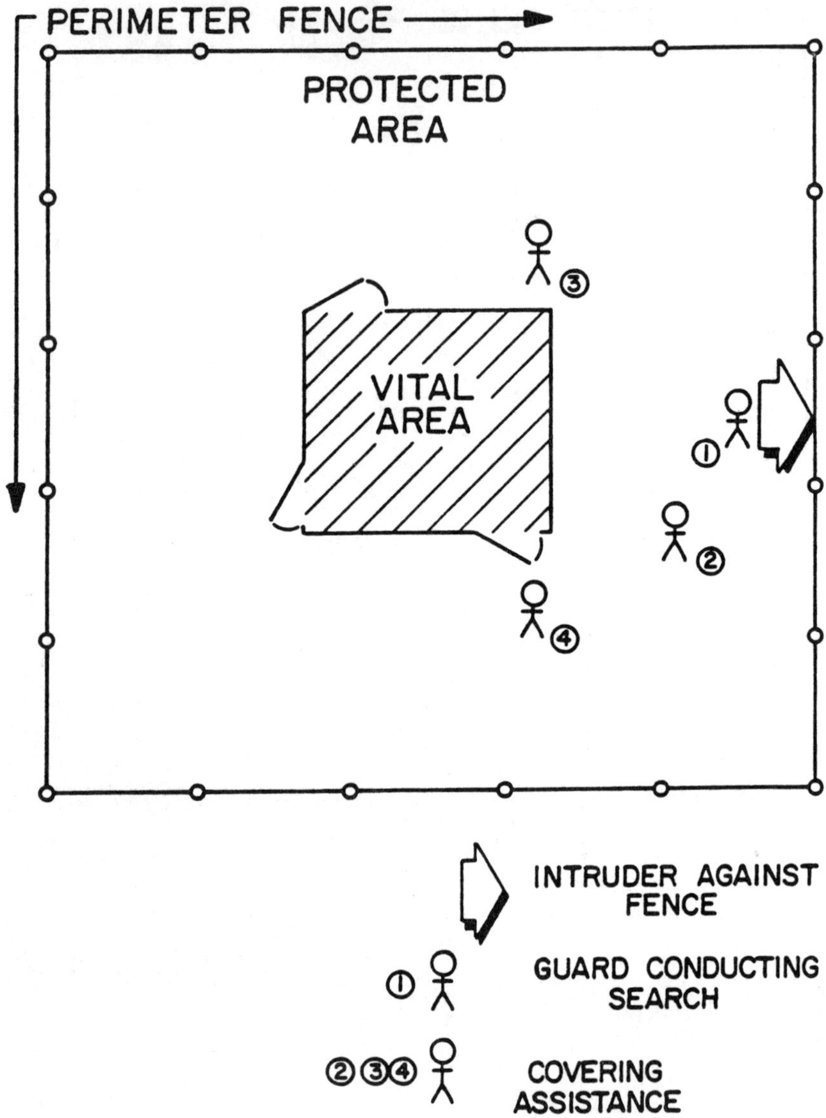

Figure 14. Guard Force Conducting Search

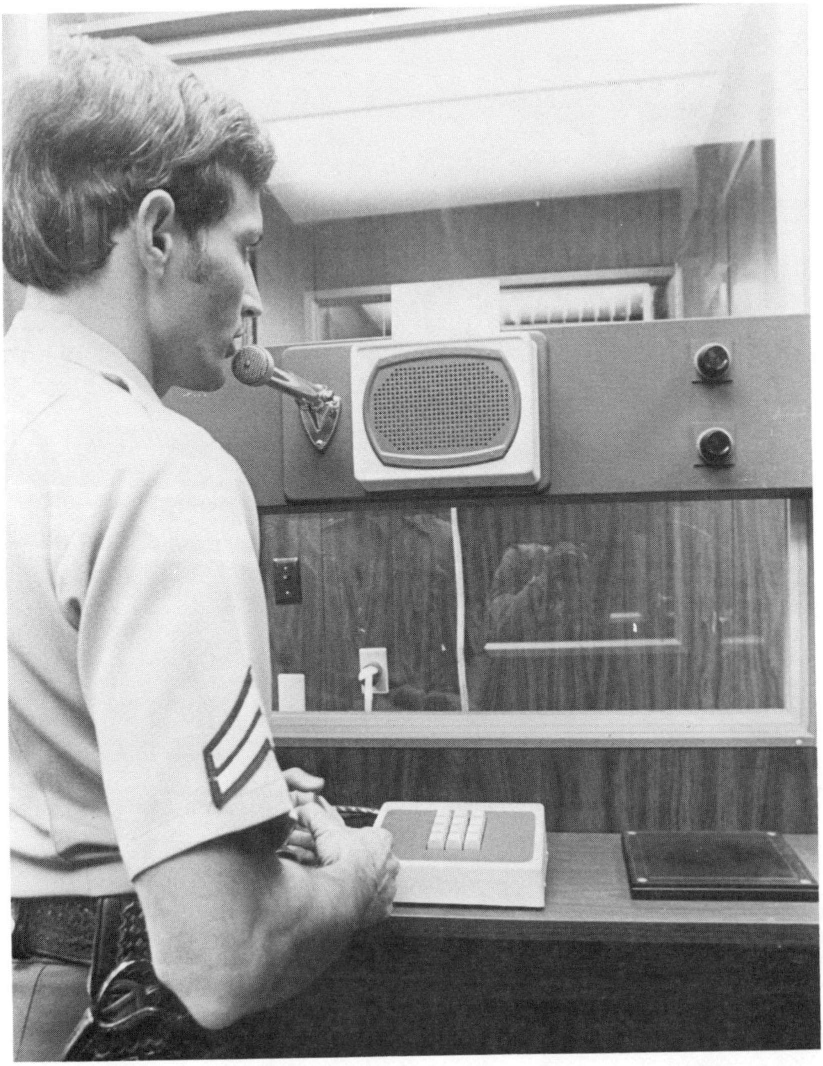

Figure 15. Inside a Bullet-Proof Guard Station. Courtesy of Allied-General Nuclear Services

Figure 16. Searching for Hidden Special Nuclear Material Using a Radiation Detector. Courtesy of Allied-General Nuclear Services.

Response	*Descriptive Title*
1.	Immediate maximum confrontation
2.	Single confrontation with a reserve
3.	Challenge from safety with a reserve
4.	Defend the vital* area
5.	Defend the vital area while lead guard confronts with a reserve (*Response 2* and *4*)
6.	Defend the vital area while lead guard challenges from safety with a reserve (*Response 3* and *4*)
7.	Defend the vital area with maximum confrontation (*Response 1* and *4*)

The advantages and disadvantages of these example tactical responses are discussed below assuming a guard force of ten or less, each guard armed and equipped with communication devices. A schematic diagram of the tactical response accompanies each discussion with the legend:

Response 1: Immediate Maximum Confrontation (three to seven guards)

This tactical response provides immediate intruder confrontation by as many guards as can be made available.

If there is another intruder in a different location(s) or if the first alarm is a feint or ruse, this technique loses effectiveness. It also causes all available guards to respond to every alarm; after a number of false alarms, the attitude of the guards may become less enthusiastic and they may be caught unprepared during an-

*For definition of vital area, see Glossary.

other so-called "routine" tactical response. In addition, if the available guards must first assemble and then confront the intruder en masse, valuable time is lost; but if the intruder is confronted piecemeal as the guards arrive, the guards may run into a fire fight or ambush.

One solution is to pre-organize each guard shift into squads of two or three where each squad has control over a sector of the

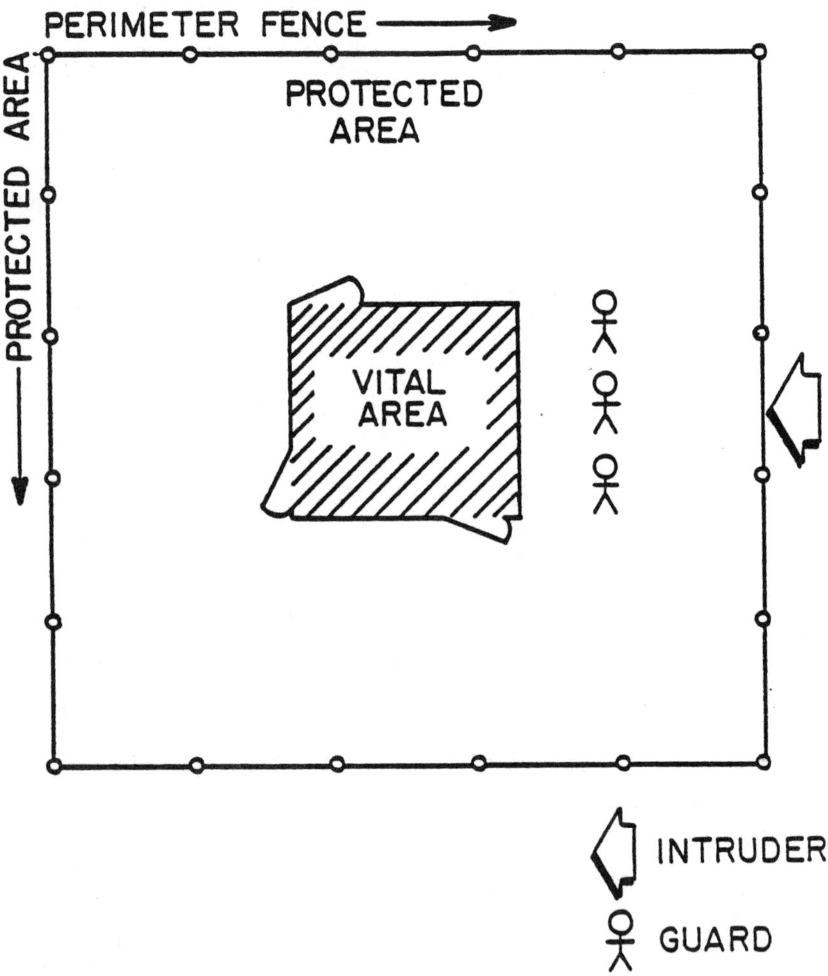

Figure 17. Immediate Maximum Confrontation

plant near its normal station. When an intrusion is discovered in a sector, the sector squad would control and organize the confrontation. The remaining squads would be called upon as required.

Response 2: Single Confrontation with a Reserve (two guards)

In this tactical response, the lead guard confronts the intruder while a second guard is held in reserve in a position of safety.

The actual severity of the intrusion is rapidly confirmed while only one guard is subjected to a possibly dangerous confron-

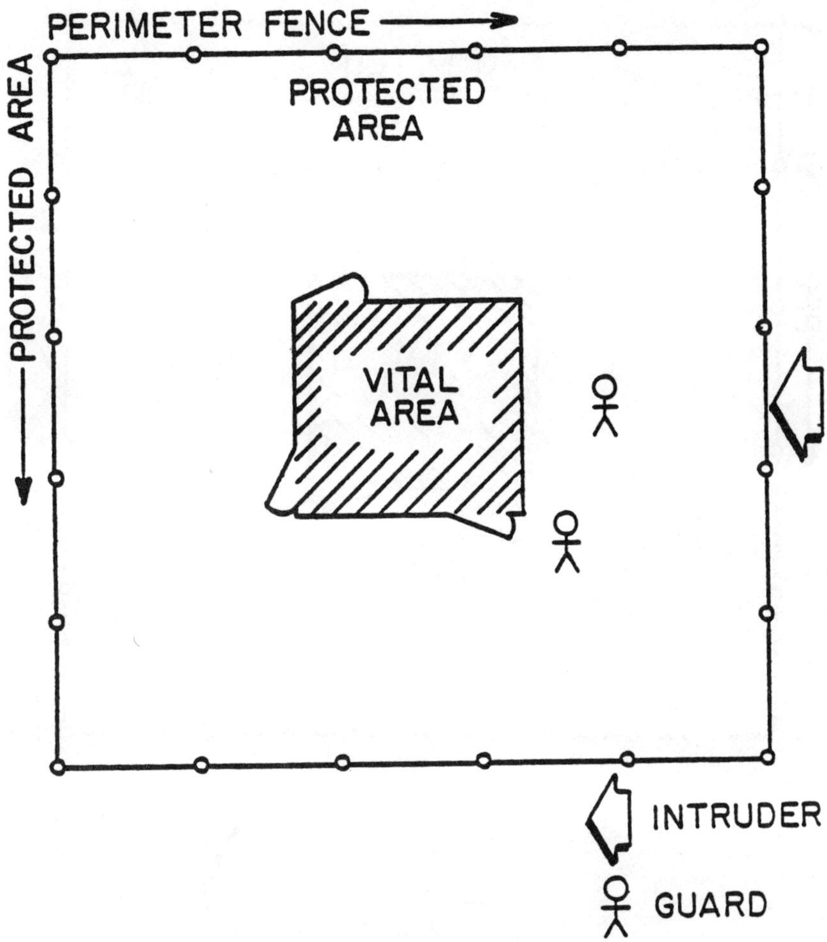

Figure 18. Single Confrontation with a Reserve

tation – one in which the guard could be killed by the intruder. If the threat has already been determined to be substantial, having a guard with a reserve verify a known danger may not save any time or slow the intruder.

Response 3: Challenge from Safety with a Reserve (two guards)

The lead guard challenges the intruder from a safe position with a second guard in reserve in a safe position.

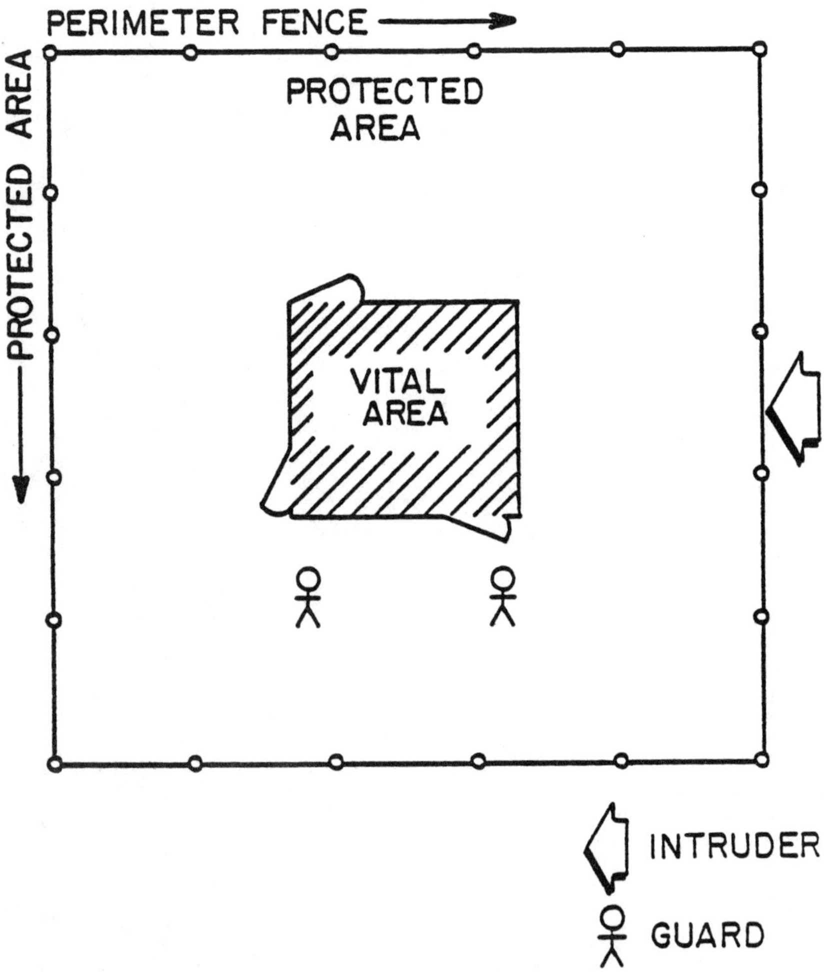

Figure 19. Challenge From Safety with a Reserve

Assuming that there are safe positions from which the guard may challenge, this tactical response allows a determination of the intent of the intruder without subjecting the guard to danger. If the intruder is noticed by an alarm or by a guard near a guard post, the same guard who participated in the threat analysis could challenge the intruder. Since the guard would be in a safe spot (the guard post), he could proceed to challenge while the reserve was enroute.

If there are no safe spots from which the guard can challenge, this tactical response is the same as *Response 2.*

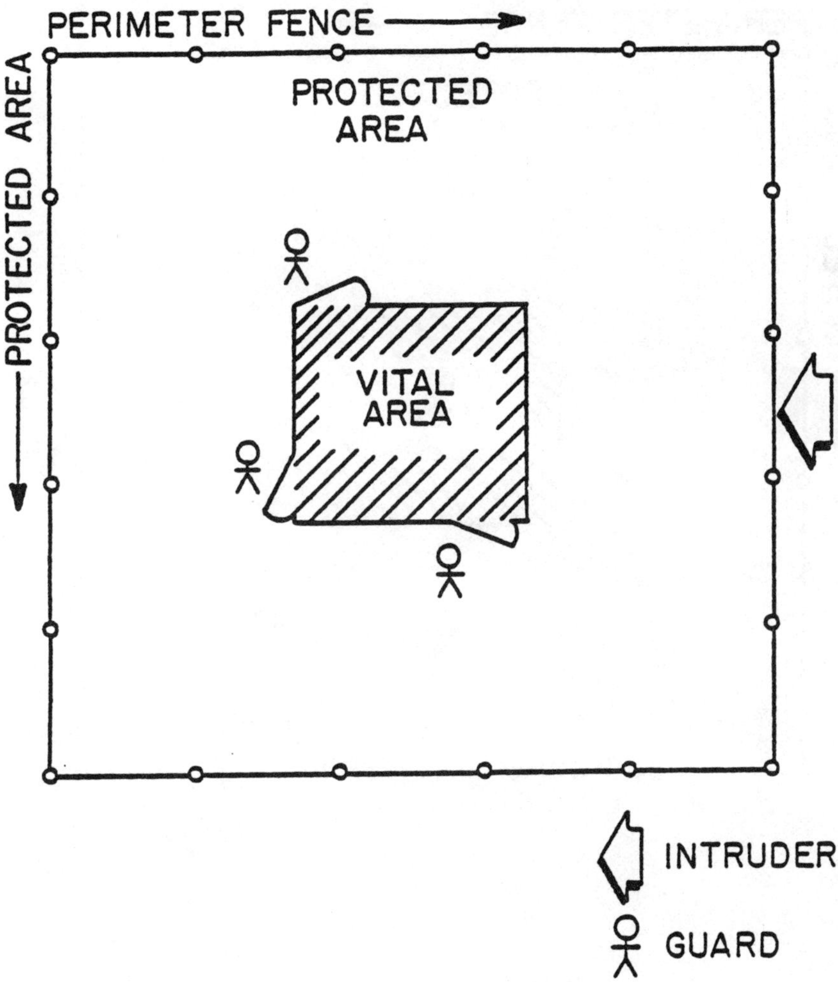

Figure 20. Defend the Vital Area

Response 4: Defend the Vital Area (one or more guards)

Response 4 consists of guarding the vital area and leaving the intruder to his own resources.

A guard is placed at each entrance to the vital area, preferably in safe positions. Although possibly unable to observe all activities of the intruder who is free to roam the protected area, each guard observes the intruder and protects the vital area until the LLEA arrives. After the LLEA arrival, the confrontation between the LLEA and the intruder is external to the vital area but probably inside the protected area.

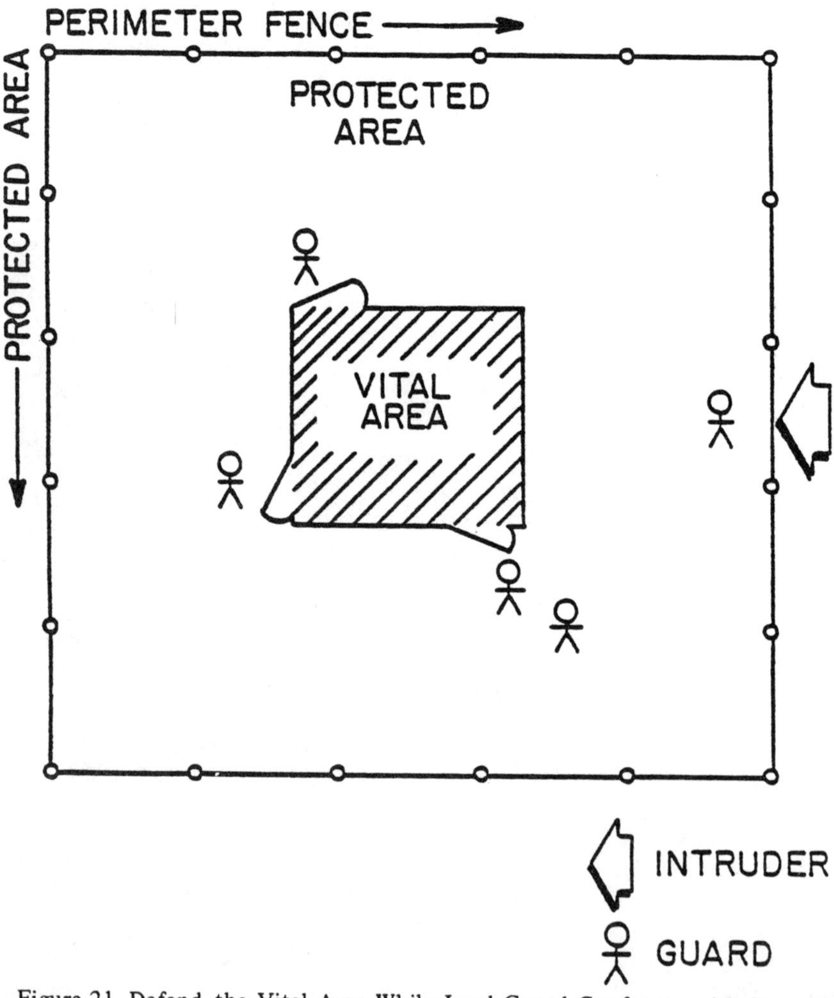

Figure 21. Defend the Vital Area While Lead Guard Confronts with Reserve

Response 5: Defend the Vital Area while Lead Guard Confronts with a Reserve (three or more guards)

A guard is placed at each entrance to the vital area while a lead guard challenges the intruder with a reserve guard looking on from a position of safety.

This combination of *Responses 2* and *4* has the advantage of having the vital area defended while the intruder is confronted, but minimizes the disadvantages of other tactical responses. The weak point of having one guard confront the intruder remains and is not overcome by guards at vital area portals. However, this tac-

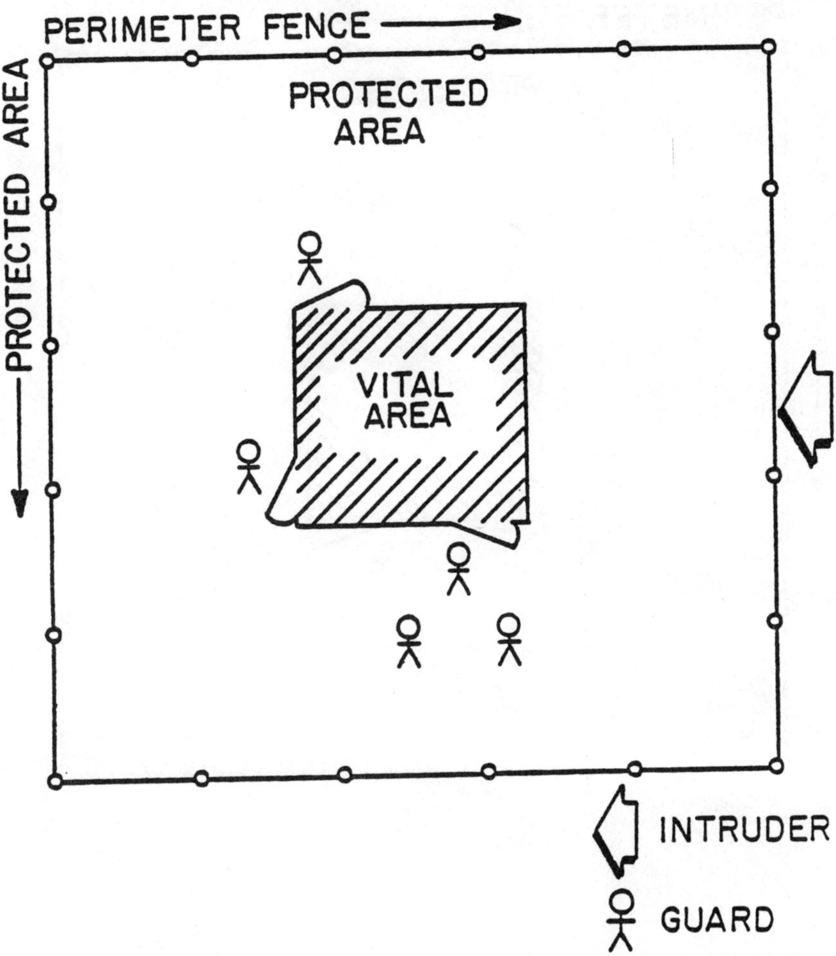

Figure 22. Defend the Vital Area While Lead Guard Challenges From Safety with a Reserve

tical response merits consideration by the guard force.

Depending upon the number of vital area portals to be defended, the required number of guards for this defense approaches the three to seven maximum available guards as shown in Figure 10.

Response 6: Defend the Vital Area while Lead Guard Challenges from Safety with a Reserve (three or more guards)

The vital area is defended by placing a guard at each entrance, in a safe position if possible. The lead guard challenges the intruder from a safe position while a second guard remains safely in reserve.

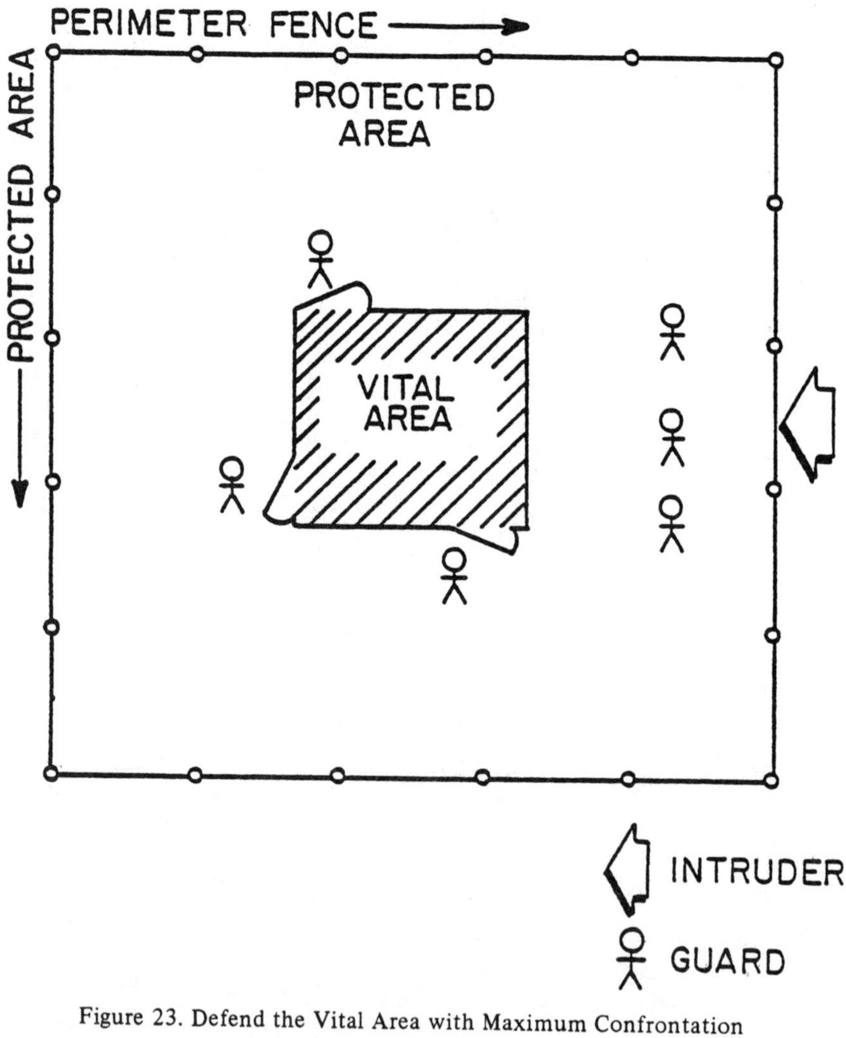

Figure 23. Defend the Vital Area with Maximum Confrontation

In this combination of *Responses 3* and *4*, the lead guard is not exposed to the intruder during the confrontation as in *Response 5*. The number of guards remains the same as for Tactical *Response 5*, but the same results are accomplished with less danger to the lead guard.

Response 7: Defend the Vital Area with Maximum Confrontation (three to seven guards)

In this combination of *Responses 1* and *4*, a guard is placed at each entrance to the vital area, preferably in a safe position. All other available guards confront the intruder.

The advantages of this tactical response are greater if there are more than ten guards on site. Since it is likely that only three to seven guards can respond and at least one of these is required to defend the plant's vital area, then only two to six guards are available to provide maximum confrontation.

If only three guards can respond, this response is not much different from *Response 5* (Defend the vital area while lead guard confronts with a reserve) with the same problems. However, if three or more guards, in addition to the guards defending the vital area portals, can respond, the intruder is faced with simple maximum confrontation as shown below in the first schematic.

Prior to any safeguard contingency each LLEA must be contacted and informed of their possible role in protecting lives and property while preserving the peace. The jurisdiction of each county sheriff's office, city police, state police, the FBI, and any other agency must be determined. The sequence of notification along with the number of officers able to respond, response time and type of equipment carried (shotguns, gas, etc.) should be made. The plant security manager should periodically invite local law enforcement officers from the agency most likely to respond with assistance to the plant to familiarize them with the plant protected area, entrances, and any special requirements. Of special concern is the need to impress upon the LLEA officers of the proper chain of command flowing from the plant management through plant security to the LLEA officers.[7] This chain of command is required to prevent the LLEA officers from overreacting and subjecting themselves, others, and particularly the plant equipment to unnecessary risk.

Procedures for contacting the various agencies should be established and rigorously followed. These contact procedures should be functionally tested periodically.

Chapter 3

SPECIFIC PROCEDURES FOR ALARMS

IN this chapter, recommendations are given for the initial reaction, alarm assessment and threat level determination, and tactical response procedure for each of the thirty-two alarms. For a number of alarms the initial reaction is to initiate an immediate guard mobilization and tactical response. In this case, the assessment and determination is carried out as a parallel effort to determine the best way for the guard mobilization to proceed.

The visual assessment list formerly developed is repeated here for convenience and is referred to hereafter as a "visual assessment":

1. number of people involved,
2. ages (adult/child), sex, and physical characteristics,
3. apparent intent or aggressiveness of entry (is it just a misunderstanding or prank, or really an armed act of aggression),
4. demeanor,
5. arms, equipment, and tools,
6. location of intruder (position relative to vital areas or vital equipment, or approximate distance to licensee's property line),
7. movement of individual (speed and apparent destination on-site or direction of travel),
8. visible badge identification,
9. vehicles,
10. degree of intoxication,

11. reaction to verbal contact and request to halt,
12. likelihood of guards ability to restrain individual without resort to use of deadly force,
13. activity in which engaged (fishing, setting explosive, etc.),
14. likelihood of malevolent acts having been committed by exiting individual.

ALARM 1: SOMEONE IN VITAL AREA – NO APPARENT AUTHORIZATION

Initial Reaction

The security supervisor initiates a guard mobilization and tactical response and requests the LLEA to respond.

Attempts to view with CCTV will be made while guards are dispatched to the area to view the individual.

Assessment and Determination

While the tactical response is being made, the security supervisor will be informed (generally without the knowledge of the unauthorized person) of the information obtained from a visual assessment of the strength of the intruder.

Tactical Response

A vital area is, by definition, an area which is susceptible to sabotage. The presence of someone in a vital area who is not recognizable as a member of the utility staff is a serious matter requiring immediate guard mobilization and tactical response while a threat analysis is made. Based upon a minimum of information, the security supervisor must choose the proper response since the guard force must react in a slightly different manner to avoid any unnecessary incidents if the apparently unauthorized individual is recognized as a member of the utility staff.

The security supervisor initiates the following responses:

Response 6 (Defend the vital area while lead guard challenges from safety with a reserve) if only one intruder has been spotted, or

Response 7 (Defend the vital area with maximum confrontation) if there is more than one intruder.

These responses are recommended because the number of guards who must enter the vital areas is limited, and, even though the vital area has been compromised, the guards at the portals are subject to immediate action if required.

The lead and reserve guard proceed immediately to the vital area. Guards at entrance portals must assume a safe position at all entrances. Any employees on the scene should attempt to disengage themselves from the apparently unauthorized individual to reduce the possibility of hostage taking. Other responses by employees may be appropriate depending upon the circumstances.*

From a position of safety, the lead guard immediately commands the individual to leave the vital area and escorts the individual to the CAS for further investigation. The reserve guard notifies the security supervisor of the confrontation, relays any important information, and covers the lead guard.

The security supervisor advises the CAS of the fact that an apparently unauthorized individual was confronted in a vital area and of the location of that area.

If the individual does not voluntarily leave the vital area, he is forcibly removed by the guards and taken into custody. Upon their arrival the LLEA is responsible for the intruder.

A search of the compromised vital area for sabotage, additional intruders, or unauthorized devices is begun immediately and is continued until complete or until the unauthorized access is explained to the satisfaction of the security supervisor. CAS is notified once the search is complete.

ALARM 2: VEHICLE FORCING ITS WAY INTO PROTECTED/VITAL AREA

Initial Reaction

The security supervisor initiates a guard mobilization and tactical response while requesting LLEA to respond.

*Having employees restrain a person would be a last-resort activity which would be justified only if the suspected individual is engaged in an obvious sabotage or theft of SNM. It would not seem reasonable, for instance, to think that employees would stand by and *only watch* another person who spread gasoline about with obvious intentions of starting a fire.

Assessment and Determination

An assessment for this alarm would seem to be unnecessary because this situation appears to be an obvious threat to the facility. However, the forceful entry might not be connected with sabotage but may be the result of a personal grudge or domestic quarrel, and thus only a police action.

After giving a general alarm requiring maximum tactical guard force response and assuming the entry is for malevolent purposes, make a threat determination based on the visual assessment of the guard or individual who originally viewed the action or from viewing CCTV.

Figure 24. Guard Station. Courtesy of Allied-General Nuclear Services.

Tactical Response

This alarm is a challenge to the ability of the security force to act rapidly and in a proper manner. The intruding vehicle prevents an accurate assessment of how many people are involved, their arms, tools, or intent. For this reason the guard force responds with maximum effort while guarding the vital areas of interest.

• The security supervisor initiates *Response 7* (Defend the vital area with maximum confrontation).

• Since a vehicle can be a rapidly moving target, it may be difficult if not impossible for the guards to arrive at the vital area entrance before the intruder. If the intruder arrives first, the guards designated to defend the vital area become confrontation guards upon arrival.

• Guards responding to a vehicle intrusion must be aware that the intruder has a portable barrier (the vehicle) that can be used to the disadvantage of the guards. The guards confronting the vehicle may have to observe the vehicle and wait for it to stop before a tactical response can be mounted. If an individual is discharged from the vehicle, he is immediately confronted.

• State laws vary but in general, the guards may not shoot at the intruding vehicle and/or its occupant until the guards have probable cause to believe a dangerous felony is either in progress or has been completed and the felon is escaping. Some states, e.g. New York, severely limit the use of deadly force by a private person against an escaping felon. Upon arrival, the LLEA assumes responsibility for the intruder.

ALARM 3: VEHICLE FORCING ITS WAY OUT OF PROTECTED/VITAL AREA

Initial Reaction

The security supervisor initiates a guard mobilization and tactical response and alerts LLEA.

Assessment and Determination

The primary concern is to determine what the individual in the vehicle left behind rather than what he intended to accomplish. While tactical response is being made, determine what the exiting individual left behind and assess the alarm:

1. notify plant operations,
2. inspect security equipment for compromise and security areas for concealed persons,
3. place guard force on alert and notify all guards, and
4. assist operations personnel in inspecting vital systems.

To deal with the exiting vehicle:

1. notify LLEA to give vehicle description: color, type, model, license number; and number of occupants, and
2. pursue vehicle to licensee's property limits with guard force members.

Tactical Response

The primary concern of the tactical response team or theft of SNM is to determine if the exiting vehicle is part of a sabotage action and, if so, whether it is the beginning, the end, or part of a continuing plan.

The security supervisor implements *Response 6* (Defend the vital area while lead guard challenges from safety with a reserve) with the lead and reserve guards using a vehicle for pursuit and as a barrier.

While enroute to defend the vital area, the guards report observations of any unusual circumstances by radio to the security supervisor. A search and inspection is made of the protected/vital areas to discover if any sabotage was accomplished or if any explosives were left behind.

The pursuing guards may have to observe and follow the intruder vehicle while directing LLEA officers to the proper location. Although intercepting and interrogating the fleeing intruder is important, the guards should refrain from using their legal right of citizens arrest unless necessary but should let the LLEA officers stop the vehicle and arrest the occupants.

ALARM 4: HARDWARE/EXPLOSIVES VIEWED (SABOTAGE AIDS)

Initial Reaction

The security supervisor initiates a guard mobilization and tactical response while requesting LLEA to respond. If appropriate, the nearest Bomb Disposal Unit (BDU) is notified and requested to assist in either search or disposal.

All security personnel are notified of the threat and instructed to follow preplanned activities. The facility control room is notified of equipment involved so that an assessment of the threat to the facility can be made. Notifying the central alarm station causes recent alarms to be reviewed and, if appropriate, guards to be dispatched to inspect for explosives in vital areas where unexplained alarms had occurred.

Assessment and Determination

While the tactical response is being made, the assessment action for this alarm is to determine the extent of the threat as viewed and if the threat is continuing elsewhere.

The person or guard viewing the hardware/explosive informs the security supervisor of his findings. The security supervisor contacts facility operations management and informs them of the discovery.

The security guard and operations personnel determine if the threat is limited to that viewed or if it is continuing elsewhere in the facility. The operations personnel inspect vital areas while security personnel inspect safeguard equipment and review alarms of the recent past few hours.

Tactical Response

This alarm consists of a guard or other utility employee viewing explosives.

The security supervisor initiates *Response 4* (Defend the vital area). This response prevents further sabotage or theft attempts unless the intruder is already in the vital area.

All employees not needed for the search are evacuated from the area with security personnel assuring an orderly exit and accounting of employees.

A search for further devices and for any unauthorized persons is performed considering the points given below. Any intruder located is turned over to the LLEA.

1. A decision of whether to shut down the plant is made prior to the search.
2. The search is conducted by police, facility security person-

nel, and *someone familiar with the normal condition of the areas to be searched.*

3. Once an area is searched and found clean, it is closed to further access.
4. Medical personnel and an ambulance are positioned nearby in case the search activities result in detonation or discharge.
5. Doors and windows are left open if possible as this may tend to vent high pressures from explosion and reduce injury from flying glass and other debris.
6. Any suspicious items are reported to the BDU and no attempt to cover, disarm, or touch the suspected item is made.
7. The use of radio transmitters and any other devices radiating even small amounts of power may have to be curtailed in certain areas until the situation is resolved.

ALARM 5: SOMEONE COMING OVER THE FENCE

Initial Reaction

The guard visually discovering the individual hurriedly places himself in a position to view the situation and relay what he views to the central alarm station. The security supervisor notifies the LLEA and then initiates guard mobilization and tactical response. If the individual is discovered via CCTV or other remote alarm, an especially dispatched security guard is sent to verify the threat.

Assessment and Determination

While the tactical response is being made and the LLEA is enroute, the guard conveys his initial observations to the CAS, then continues to observe from a safe position. Because of the urgent nature of this alarm, the involvement of the security supervisor in assessment and determination may be minimal. Threat level determination is made by the guard at the scene or, if time allows, by the security supervisor based on the information obtained from the visual assessment.

Tactical Response

Observing someone climbing the perimeter fence is a serious

matter. However, it is easy to think of many situations in which the person climbing the fence is an innocent individual who has no malevolent intent; e.g. a hunter coming after his prize hunting dog which managed to crawl under the fence, or a fisherman who wants to fish in the warm effluent of the plant.

The guard response must be adequate to treat the possibly innocent intruder with firmness and fairness without raising the risk that a true malevolent intrusion is misinterpreted or subjects the guards to unnecessary exposure. In all cases the intruder is ultimately taken into custody and escorted to the central alarm station. Even a plant employee, including the plant manager, who is found climbing the fence is taken into custody and required to explain his conduct to the satisfaction of the security supervisor and superiors.

This alarm is initiated by a visual sighting, an Omni-alarm, or a fence motion detector and requires an immediate mobilization while the threat analysis is being made. While the safety of the plant is the major concern, the guard taking the individual into custody must do so with a minimum risk of danger to himself and with some regard for the personal dignity of the intruder.

As soon as possible, a reserve guard is sent to assist the guard who visually established the presence of an intruder.

The security supervisor considers the information obtained from the visual assessment and implements:

Response 2. (Single confrontation with a reserve) if there is a single intruder without apparent malevolent intent and without apparent means to carry out a violent assault, or

Response 7. (Defend the vital area with maximum confrontation) if more than one individual is observed penetrating the perimeter fence or if there is visual evidence of apparent means to carry out a violent assault.

Given below are specific actions for the response guards to take while approaching the intruder but prior to confrontation.

1. From a reasonably safe vantage point, the lead guard establishes verbal contact with the intruder noting the conduct of the intruder as a result of the verbal challenge. By this time, the intruder should realize that he should not be climbing the fence.

2. If the intruder is already inside the fence, command him to

halt. A "halt or I will shoot" command is probably legally permissible, particularly if the guard does not draw his weapon at that time. The threat to use deadly force without an intention to carry out the threat is not the use of deadly force. This action is probably at most an assault and may be justified under these circumstances. At this point, the intruder either stops or keeps moving and may have malevolent or violent intentions. The observation of the guard must immediately determine the apparent intent of the intruder.

If after having made the assessment it is determined that the intruder probably is not armed, the lead guard advances and confronts the intruder at close range. Other guards view this movement from a safe vantage point. At close range the lead guard then reassesses the intruder and determines if any changes have occurred in the original assessment.

1. If the intruder is cooperative and without violent intentions, he is relieved of any possessions that could be used as a weapon, and then escorted to the central alarm station under private persons arrest.

2. If the intruder has any apparent capability to commit or any apparent intention to engage in sabotage, he is taken into custody using only that force needed to overcome the force exerted by the intruder. In most states a person being arrested by a private party (a guard in this case) is under no compulsion to submit to that arrest and may legally resist any arrest he believes is improper. The guards must be aware that deadly force is never justified for a misdemeanor and most (but not all) states would regard a simple trespass as a misdemeanor.

3. If the intruder resorts to the use of deadly force, the reserve guards in protected places return the fire with the intention of neutralizing the intruder.

4. The intruder is delivered to responding LLEA officers. However, the guard or facility is not relieved from a false arrest charge since the LLEA merely transports the intruder under authority of the private person's arrest.

If all the guards are in protected positions and if LLEA personnel have arrived and can be brought to bear on the confrontation without undue risk to the LLEA officers, the guard force defers to the LLEA officers. However, this is possible only if the intruders are contained and are not able to continue their penetration or apparent mission.

A report is written immediately. Although probably not admissible as evidence, it serves to refresh the guards' memory should legal action require testimony.

ALARM 6: SOMEONE GOING OVER FENCE AWAY FROM PLANT

Initial Reaction

The guard visually discovering the individual hurriedly places himself in a position to view the situation and relay what he views to the central alarm station. The security supervisor notifies the LLEA and then initiates guard mobilization and tactical response. If the individual is discovered via CCTV or other remote alarm, an especially dispatched security guard is sent to verify the threat.

Assessment and Determination

While the tactical response is being made, the guard immediately conveys to the security force supervisor enough information to allow the supervisor to initiate a protective search/inspection of the reactor and also to apprehend the fleeing individual. Initiate protective search/inspection of the reactor to include:

1. notify plant management of details of observation,
2. assist operations personnel in plant search and inspection,
3. initiate patrol and check out all facility safeguard equipment, and
4. review recent alarms for possible indication of location of sabotage action.

Concurrent with protective search/inspection, initiate apprehension of fleeing individual by executing the following:

1. Assuming exiting individual halts and stands fast as commanded, notify the security supervisor and provide the following information from which a threat determination can be made:
   (a) was apprehension inside or outside the fence,
   (b) location of apprehension,
   (c) number of individuals intercepted,

(d) apparent intent and demeanor of individual intercepted,
(e) statements made by fleeing individual,
(f) apparent arms,
(g) an assessment, based on guards impression, of likelihood of malevolent acts having been committed by individual intercepted.

2. Assuming exiting individual did not stop when verbally commanded to do so by the guard, the security supervisor is to make a threat determination based on the guard's visual assessment.

Tactical Response

The primary aim of the guard force is not to apprehend the fleeing individual unless he has stolen a quantity of Special Nuclear Material; however, if apprehension is reasonable, the individual is stopped in hopes that information concerning his activity inside the plant could be obtained. At this time, a number of guards will be required to inspect the protected area for any remaining intruders and to be available in case additional intruders are found.

To conserve the response capacity of the guard force the security supervisor uses *Response 2* (Single confrontation with a reserve). This response used in conjunction with a vehicle allows a quick response with a minimum number of guards.

When intercepted, the fleeing individual is apprehended by the felony arrest procedures commonly used by police. These procedures are required because the two guards will probably be making contact with the fleeing individual in a remote area without the benefit of protection from other guards or support of LLEA officers.

The intercepted individual is interrogated to determine what activity he was engaged in. Any information obtained is radioed immediately to the security supervisor.

When the LLEA arrives the suspect is turned over to LLEA custody and charges are made that are consistent with local and state laws.

ALARM 7: COMPONENT DAMAGE DETECTED (ALARM AND/OR OPERATION)

Initial Reaction

The person discovering the damage informs the central alarm station. The security supervisor dispatches a guard to the scene of damage, then immediately produces records of recent alarms and other information needed to determine if any potential threats are in progress.

The security supervisor initiates a guard mobilization and tactical response and notifies the LLEA.

Assessment and Determination

Alarm assessment consists of the security supervisor determining whether the damage is a result of sabotage and, if so, whether the sabotage is still underway elsewhere or is just limited to that already discovered. Operations personnel assist in assessing the nature and extent of the damage and determine security significance.

While a tactical response is being made, the person or guard viewing the damage contacts the security supervisor and informs him of the damage. The security supervisor relays the information to reactor operation management and queries them about other equipment failures that may be part of a sabotage plan being pursued. Delegated members of the guard force and operations staff make visual inspection of vital equipment and vital equipment access control points.

Tactical Response

This alarm consists of a guard or other facility employee finding component damage which appears to possibly be the result of sabotage. A tactical response is requested and executed before the damage can be evaluated to determine if it was actually caused by sabotage. The object of the tactical response is to interrupt sabotage action that is taking place and prevent any sabotage action not yet initiated.

The security supervisor initiates *Response 4* (Defend the vital area). This response prevents further sabotage attempts unless the saboteur is already in the vital area.

Operations personnel, accompanied by a guard, makes a rapid survey of the vital area to locate further sabotage and any unauthorized persons.

Upon arrival the LLEA is responsible for the removal of any intruders located.

ALARM 8: SOUND OF EXPLOSION

Initial Reaction

Guard mobilization and tactical response is made immediately so that the sound can be evaluated and the activity assessed. LLEA is notified.

Assessment and Determination

This alarm requires the sound of an explosion to be investigated to determine if the explosion is connected with facility sabotage. The security supervisor deploys all available guards to points throughout the reactor to inspect for damage and calls the control room informing them of the explosion and asking if they have any indication of equipment failure.

If the explosion is located at the reactor site, the discoverer informs the security supervisor of the location and begins a search for individuals preparing to commit further sabotage.

Tactical Response

This sound could result from an industrial accident or it could signal an undetected sabotage action in progress.

The tactical response is a combination of limiting further sabotage (if actually in progress), discovering the source of the explosion, and assisting injured personnel or personnel evacuating the area.

The security supervisor initiates *Response 7* (Defend the vital area with maximum confrontation) while he and the guards remain aware that the explosion may be a diversion to draw attention away from the real attack.

The designated confrontation guards search for the source of the explosion while maintaining a state of alertness for other suspicious activity.

ALARM 9: DOOR ALARM (CERTAIN DOORS)

Initial Reaction

Certain unexplained door openings in the vital area demand that the security supervisor initiate a guard mobilization and tactical response by the guard force such that two or more guards are available at the threatened area in the minimum time. In addition, the LLEA is asked to respond.

Assessment and Determination

Being inherently reliable, door alarms are indications of actual intrusions. The guard force's responsibility is to determine if the intrusion is authorized and thus the alarm is simply the result of administrative failure or if there is an actual intrusion. In this alarm the assessment and the tactical response occur simultaneously.

Tactical Response

Certain doors in or leading to vital areas are the last alarmed indication that unauthorized persons are present. When an alarm indicating that a vital area has been entered is received, a rapid response without a threat analysis is required.

A response requiring that the vital area be protected assures that other entrances are not compromised and provides a mechanism to intercept anyone attempting to escape through another exit. In addition, having guards at all vital area exits assures additional quick response should the need arise. However, the guards may not normally be allowed to enter the area and may not have the capability of unlocking the portal without obtaining special keys, key cards, and permission. If two guards on each shift are trained and cleared to enter vital areas in emergencies, the security level is considerably increased.

The security supervisor chooses a response based on information obtained without an on-the-scene threat analysis. If the door alarm is to a small vital area which is physically separated from other vital areas, *Response 5* (Defend the vital area while lead guard confronts with reserve) is quick since only two guards are required to respond. In addition, this response avoids the confusion of placing a number of guards in close quarters where they may be in each other's way.

If the vital area is all one big area, as it often is in older facilities, *Response 7* (Defend the vital area with maximum confrontation) is recommended based on the current practice of most utilities in allowing guards in vital areas (some in control rooms).

ALARM 10: PERSONNEL DETECTOR ALARM

Initial Reaction

The security supervisor initiates a guard mobilization and tactical response while requesting LLEA to respond.

Assessment and Determination

This alarm is a back-up to door alarms and indicates the presence of personnel in vital areas. An alarm of this type means either that the primary alarms failed to work and nonauthorized personnel are in vital areas or it is a false alarm.

After the first guard tactical response team has arrived they provide the security supervisor with a visual assessment.

Tactical Response

Since this alarm detects the presence of people inside the vital area, the same comments given for Alarm 9 (Door alarm) are appropriate.

ALARM 11: FLOOR WEIGHT DETECTOR ALARM (CERTAIN PLACES)

Initial Reaction

The security supervisor initiates guard mobilization and tactical response while requesting LLEA to respond.

Assessment and Determination

Since this alarm would probably be interior to other alarms and be inside a vital area, it is regarded as an indication of intrusion. This alarm may indicate that a serious threat to vital equipment is in progress after successful evasion of barrier alarms or it may be a false alarm. Assessment occurs simultaneously with the tactical response.

Tactical Response

A floor weight detector is used inside a vital area as the final opportunity to detect an intrusion; therefore, the same response given in Alarm 9 (Door alarm) is applicable.

ALARM 12: GUARD-REQUESTS-AID ALARM

Initial Reaction

Guard mobilization and tactical response by the guard force is such that multiple guards are dispatched to the area where the aid is requested. LLEA is notified.

Assessment and Determination

This alarm requires immediate maximum tactical response without waiting for an assessment and determination. Concurrent with the guard mobilization, a guard located near the guard giving the duress signal is directed to approach the area of concern, make a visual assessment, and give aid to the guard signaling duress.

Tactical Response

Because of the unknown nature of this alarm, *Response 7* (Defend the vital area with maximum confrontation) is initiated. An alarm of this nature is reserved for a bonafide security emergency so that the tactical response to this alarm is the same as if there were a security alert.

ALARM 13: AREA RADIATION ALARM

Initial Reaction

A radiation alarm from some part of the plant may be evidence of sabotage in process or completed. A guard accompanied by a radiation monitor expert is dispatched to view the area from where the alarm sounded. All other guards are alerted about the alarm and instructed to be specially alert to possible sabotage or theft activities. LLEA is notified.

Assessment and Determination

This alarm alerts the plant personnel of the presence of excessive radiation in the area. The guard force is deployed throughout the plant to determine if the alarm is an indication of sabotage or theft having been completed or in progress, or is a false alarm.

Tactical Response

This alarm, like Alarm 8 (Sound of explosion), could result from an accident, sabotage, or malfunction.

The tactical response is a combination of limiting further sabotage (if actually in progress), discovering the cause of the alarm, and assisting injured personnel or personnel evacuating the area.

The security supervisor implements *Response 7* (Defend the vital area with maximum confrontation) while the entire guard force remains aware that the alarm may be a diversion.

Some of the confrontation guards patrol other areas of the plant observing any suspicious activity, escort and direct personnel, and assist with security matters.

ALARM 14: SOMEONE COMING TOWARD OR NEAR FENCE

Initial Reaction

A guard is dispatched to visually assess the situation from a safe vantage point. Attempts to view the situation via CCTV are made while the dispatched guard is enroute.

Assessment and Determination

From a position of safety, the guard relays the visual assessment information to the security supervisor who then determines the likelihood of any malevolent action and what additional guard action, if any, is required.

Tactical Response

The response is made with the benefit of the information obtained during the assessment; however, the actual situation as to the intent of the potential intruder, and whether or not there were other intruders elsewhere inside or near the plant is not known by the responding guards. Even though the intruder is outside the protected area he can be guilty of trespass, which in some states, e.g. California, is not a criminal offense, or of some weapons/explosives possession offense. The action of the security supervisor must take into account the potential criminality of the intruder's action, as well as the recent security information.

Since the observed individual is only a potential intruder, *Response 3* (Challenge from safety with a reserve) is initiated inside the protected area while two additional guards are dispatched outside the protected area to make immediate contact with the trespasser via *Response 2* (Single confrontation with a reserve) or *Response 3*, depending upon the availability of a vehicle.

As soon as possible the guards confronting the trespasser outside the protected area establish verbal contact with the trespasser, informing him that he is trespassing and is subject to the local penalties. If a vehicle is available, contact is established from behind the open car doors as is commonly done by peace officers making a felony arrest. The conduct of the trespasser in response to the verbal contact is carefully observed.

The guards make a visual assessment of the likelihood that the trespasser is armed, taking into consideration the opinion of the guard involved in the initial reaction and the threat analysis.

The trespasser is told to stand fast so that the guard can approach and ask a few questions.

Depending upon the discretion of the security supervisor, the trespasser is detained and later released to the custody of the LL-EA officers, or released with an admonishment.

ALARM 15: SOMEONE IN PROTECTED AREA – NO APPARENT AUTHORIZATION

Initial Reaction

A guard is dispatched to view and evaluate the individual.

Assessment and Determination

The security supervisor (generally without the knowledge of the unauthorized person) is notified of the situation, and the guard is requested to observe from a safe distance and inform the security supervisor of the visual assessment.

The security supervisor makes a threat level determination based on the information obtained.

Tactical Response

This alarm is a situation requiring some tact and caution by the guard since the apparently unauthorized individual could be a member of a team bent on sabotage or theft or an innocent individual with no malevolent intent.

If the individual is recognized as a member of the facility staff then the guard force can be less cautious in their approach; however, more tact would be required to avoid unnecessarily offending an employee.

If it is determined that the unauthorized person represents a threat sufficient to cause a guard mobilization and tactical response, the security supervisor implements *Response 5* (Defend the vital area while lead guard confronts with a reserve) or, if there is evidence that the intruder is armed or intends malevolent action, *Response 7* (Defend the vital area with maximum confrontation).

The guard discovering the intruder keeps the apparently unauthorized individual under surveillance until the guard force can respond. The guard or other employee notifying the security supervisor identifies himself so that he may be contacted by guards for further information if required.

A reserve guard is dispatched to the area. The lead guard then approaches the individual while the reserve guard observes from a position of safety.

The apparently unauthorized individual is informed that he does not have the proper badges displayed and is allowed to produce a proper badge. If proper badges and credentials are not produced or are not to the guard's satisfaction, the individual is escorted back to the central alarm station for further investigation.

Any equipment, tools, and materials in possession of or near the individual are searched for contraband prior to escorting the individual back to the central alarm station.

If, at the CAS, the individual still cannot produce the proper badge, other guard stations are informed that an unauthorized individual is in custody and to be especially alert for suspicious activity.

ALARM 16: SOMEONE GOING AWAY FROM FENCE

Initial Reaction

A guard is dispatched to a position inside the protected area where verbal contact can be made with the individual moving away from the fence. The central alarm station dispatches a guard in a vehicle to intercept the individual.

Assessment and Determination

The security supervisor is notified and is provided with the guard's visual assessment information. The security supervisor assesses the situation and alerts the security guard in the vehicle in order to make him as knowledgeable as possible about the situation. The threat level determination involves a determination of what has already been accomplished and not what the future threat might be. Operations supervision is alerted to the possibility of sabotage having been committed.

Tactical Response

The primary concern is to determine what, if anything, has been accomplished by the trespasser, and if he is part of a sabotage or theft activity currently in progress.

The response is the same as for *Alarm 14* (Someone coming toward or near fence) unless the threat analysis indicated that the person seen near the fence was more than a mere trespasser. The tactical response would then be the same as for *Alarm 6* (Someone going over fence away from plant) except that the guard may elect not to use the felony arrest technique if, in his opinion, it is not warranted by the apparent circumstances.

ALARM 17: SUSPICIOUS DEMEANOR VIEWED

Initial Reaction

A guard is dispatched to view (from a safe place) the actions of the suspect individual.

Figure 25. Checking Identification of Vehicle Operator. Courtesy of Allied-General Nuclear Services.

Assessment and Determination

Suspicious demeanor is an action that causes a reasonable person to believe something abnormal is occurring. To assess the alarm a guard must regard the situation as potentially a sabotage action in progress while realizing it may be an innocent situation or something less than a sabotage action.

The guard or other employee whose suspicions have been aroused contacts the central alarm station and discusses the reason for the suspicion with the security supervisor.

A reserve guard is dispatched to view the scene. He remains at a safe distance while in continuous contact with the CAS and while observing the lead guard.

The lead guard advances, states his authority, and then commands the suspected individual to stand fast and to keep his hands in view. The individual is then requested to explain his presence, intentions, and authority for entrance.

The information obtained is relayed to the CAS where a determination of the threat is made.

Tactical Response

A tactical response for this alarm is requested only after the two guards in the area have confronted the suspicious individual and have supplied the security supervisor with enough information to cause him to decide that a tactical response is required.

The security supervisor initiates *Response 7* (Defend the vital area with maximum confrontation).

The response is carefully executed with certain guards designated for confrontation and other guards for reserve. The reserve guards attempt to observe the confrontation from a protected vantage point.

Other guards begin search techniques to determine if any other suspicious or unauthorized activities are in progress.

ALARM 18: A BARRIER BREACHED – HOLE IN FENCE/DOOR OPEN

Initial Reaction

All guards are deployed throughout the facility to observe and report from positions of safety. The central alarm station is notified.

Assessment and Determination

Assessing the alarm represented by a discovery of a breached barrier requires the guard force to determine if:

1. sabotage or theft has occurred and is still being accomplished, or
2. sabotage or theft has occurred and is no longer being pursued, or
3. sabotage or theft is about to occur, or
4. there is some explanation for the breached barrier.

The primary concern of the guard force is to prevent sabotage or theft hence the first three items are of urgent interest. The fourth item is of interest because it is mutually exclusive of the others.

To determine the threat represented by a breached barrier the security supervisor:

1. immediately informs the operations shift supervisor of the breached barrier,
2. checks the deployment of guards while informing guard personnel and other operating personnel of the likelihood of an intruder being present,
3. inspects all safeguard equipment while operation personnel inspect vital areas,
4. reviews past alarms for a clue as to the possible location of an intruder, and
5. establishes roving patrols of deployed guards in an effort to locate any intruders.

Tactical Response

A tactical response is made only if the threat analysis has provided information sufficient to raise suspicion that sabotage or theft action was possible or taking place. Since the assessment might not provide enough data to provide a focal point for the response, the vital area is protected while searching for intruders or additional information.

The security supervisor initiates *Response 4* (Defend the vital area) and prevents the guards from going first to the hole in the fence or to the open door and then to the assigned post.

Guards not assigned to vital area portals perform a search and inspection of the entire plant. Since the alarm that initiated this response did not include sighting of an individual, the guard force must be prepared to encounter any sort of intruder. The search and inspection are performed in an organized manner with guards aware of the proximity of other guards. The failure to properly organize might mean that guards would surprise and possibly shoot other guards also involved in the search.

ALARM 19: OMNI-DETECTOR ALARM

Initial Reaction

View with CCTV or send a guard to view the area. Depending upon the weather, the frequency of false alarms to real alarms is probably low.

Assessment and Determination

The assessment involves determining if there was an intrusion and, if so, how many people are involved. To make this determination the security supervisor must view the area with CCTV looking for intruders, as well as dispatch a guard to view the area from which the alarm was received; the guard is to look for causes of the alarm such as snow or birds.

If intruders are discovered, the security supervisor relays the guard's visual assessment information to the tactical response team who are to respond to the intrusion.

Tactical Response

A threat analysis is made before a guard mobilization and tactical response is requested. The Omni-detector sounds an alarm if a person is detected. The threat analysis is presumably done fast enough to visually detect the number of persons involved as well as their location.

The choice of tactical response depends heavily upon the information obtained from the assessment and determination. However, the most likely choice is *Response 7* (Defend the vital area with maximum confrontation) or *Response 5* (Defend the vital area while a single guard confronts with a reserve).

ALARM 20: WATER INTAKE DETECTOR ALARM

Initial Reaction

View with CCTV or send guard to view area. Ratio of false alarms to real alarms is probably high; however, the sensitivity of the facility to blocked water intake would dictate personal inspection by a guard.

Assessment and Determination

To determine the threat represented by this alarm, a member of the security force must view the area for the presence of an unauthorized individual, boat, or equipment. The plant control room is alerted to the fact that an alarm was received at the water intake and that the alarm is being investigated. Should an unauthorized individual, boat, or equipment be discovered at the water intake, the guard relays visual assessment information to the security supervisor.

Tactical Response

A guard mobilization and tactical response is required only after a threat analysis has indicated a response was necessary.

The same comments given in Alarm 19 (Omni-detector alarm) apply to this alarm.

ALARM 21: CCTV MOTION DETECTOR ALARM

Initial Reaction

This alarm indicates a pattern of movement in an area monitored by a special CCTV.

The first guard reaction is to view the CCTV from the central alarm station to determine if there is a reasonable explanation for the alarm. As a follow-up or if viewing the CCTV is not productive, a guard posted in the area covered by the CCTV is routed to the area of concern to observe.

Assessment and Determination

The level of threat will be determined by the security supervisor based on the information obtained from viewing the CCTV giving the alarm and from the visual assessment supplied to him by the guard viewing the scene.

Tactical Response

A guard mobilization and tactical response is required only after a threat analysis has determined a response was necessary.

Based on the threat analysis, the security supervisor chooses the proper response using the same rationale as given in Alarm 19

(Omni-detector alarm).

ALARM 22: AUDIO LISTENING DEVICES ALARM

Initial Reaction

These devices are generally located in quiet, remote places around the perimeter of the plant. The sounds of voices or vehicles indicate that people are approaching the fence.

A guard is dispatched to the area of the alarm that is inside the perimeter fence; he surreptitiously observes the activity in the area.

Assessment and Determination

Since these devices are around the exterior of the protected area, they do not represent an intrusion. From a safe vantage point, the guard observes and relays his visual assessment information to the security force supervisor.

The threat determination is made by the security supervisor based on the visual assessment information.

Tactical Response

A threat analysis is required before a guard mobilization and tactical response is initiated.

The comments given in Alarm 19 (Omni-detector alarm) pertain here.

ALARM 23: METAL/EXPLOSIVE DETECTOR ALARM

Initial Reaction

The initial reaction is to provide a guard who, from a safe distance, views the corrective action taken by the person manning the metal/explosive detector.

Assessment and Determination

The alarm is at a guard monitored portal and is probably a routine matter to correct. However, there must be a second guard in the vicinity to cover the first guard in case the metal/explosive detector uncovers a covert attempt, which then erupts into an overt attempt of sabotage or theft of SNM.

There is no threat determination unless weapons, incendiary devices, or explosives are discovered.

Tactical Response

This alarm would most probably occur at the portal detector at the entrance to the plant.

Because of the limitation on the mobility of the person causing the alarm, the action of the guards is to provide *Response 2* (Single confrontation with a reserve). Since there probably is a metal/explosive detector operator and one guard already in the area, this brings the number of armed response individuals in the area to three with one unarmed detector operator available to assist the armed individuals.

The original guard in the area performs a patdown search. The patdown search is permissible and not an invasion of privacy because the metal/explosive detector provides the guard with probable cause to believe that a crime is being or is about to be committed.

ALARM 24: BOMB/EXTORTION THREAT

Initial Reaction

Initial reaction to the threat posed by a bomb or extortion threat is a management/security activity.

The switchboard operator or other receiver of a telephoned bomb threat follows recommendations contained in the following form Figure 26, *Telephone Procedures Bomb Threat Checklist,* or similar procedures.[8]

The supervisory staff of the facility is notified of the threat. If the threat information received indicates an explosion is imminent, the plant is evacuated.

Assessment and Determination

Substantial literature on procedures for responding to bomb threats and similar extortion threats exist in the law enforcement area. The scope of various plans ranges from decisions and actions to be made by top management down to actions even for facility

TELEPHONE PROCEDURES
BOMB THREAT CHECKLIST

INSTRUCTIONS: BE CALM. BE COURTEOUS. LISTEN, DO NOT INTERRUPT THE CALLER. NOTIFY SUPERVISOR/SECURITY OFFICER BY PRE-ARRANGED SIGNAL WHILE CALLER IS ON LINE.

NAME OF OPERATOR TIME DATE

CALLER'S IDENTITY
SEX: __Male __Female __Adult __Juvenile APPROXIMATE AGE: Years ____

ORIGIN OF CALL

__Local __Long Distance __Booth __Internal (From within bldg?) If internal, leave plug in board.

VOICE CHARACTERISTICS		SPEECH		LANGUAGE	
__Loud	__Soft	__Fast	__Slow	__Excellent	__Good
__High Pitch	__Deep	__Distinct	__Distorted	__Fair	__Poor
__Raspy	__Pleasant	__Stutter	__Nasal	__Foul	
__Intoxicated	__ ______ other	__Slurred	__Lisp	__ ______ other	
			__ ______ other		

ACCENT		MANNER		BACKGROUND NOISES	
__Local	__Not Local	__Calm	__Angry	__Factory Machines	__Trains
__Foreign	__Region	__Rational	__Irrational	__Bedlam	__Music
__Race		__Coherent	__Incoherent	__Animals	__Quiet
		__Deliberate	__Emotional	__Office Machines	__Voices
		__Righteous	__Laughing	__Airplanes	__Mixed
				__Street Traffic	
				__Party Atmosphere	

BOMB FACTS

PRETEND DIFFICULTY WITH HEARING
KEEP CALLER TALKING
IF CALLER SEEMS AGREEABLE TO FURTHER CONVERSATIONS, ASK QUESTIONS LIKE: WHEN WILL IT GO OFF? Certain Hour ______ Time Remaining ____
WHERE IS IT LOCATED: Building ________ Area ________
WHAT KIND OF BOMB? WHERE ARE YOU NOW? ________
HOW DO YOU KNOW SO MUCH ABOUT THE BOMB? ________
WHAT IS YOUR NAME AND ADDRESS? ________
IF BUILDING IS OCCUPIED, INFORM CALLER THAT DETONATION COULD CAUSE INJURY OR DEATH. Did caller appear familiar with plant or building by his description of the bomb location? Write out the message in its entirety and any other comments on a separate sheet of paper and attach to this checklist.

ACTION TO TAKE IMMEDIATELY AFTER CALL

Notify your supervisor/security officer as instructed. Talk to no one other than instructed by your supervisor/security officer.

Figure 26

visitors. The scope of what follows here is directed at activities of the security organization for a nuclear facility.

Upon receipt of a bomb/extortion threat, whether by telephone* or other means, the facility management determines if the threat warrants concern. The assessment includes collecting and assessing all available information, determining if the threat is serious, and notifying the FBI and NRC of the alarm and of the threat level determination.

Tactical Response

If the threat is determined to be serious, security force actions include the following:

1. If part of the facility plan, a signal is sounded throughout the plant.[9]
2. All security personnel are notified of the threat and are instructed to follow preplanned activities.
3. The nearest Bomb Disposal Unit (BDU) is notified and assistance is requested for either search or disposal support.
4. If evacuation of the plant (full or partial) is required, security personnel assist as appropriate in assuring orderly exit and accounting of personnel.
5. In case a search for the bomb is initiated, the following points are considered:
   (a) A decision of whether to shut down the plant is made prior to the search.
   (b) The search is conducted by police, facility security personnel, and *someone familiar with the normal condition of the areas to be searched.*

---

*In regard to threats received by telephone, there are several points that merit consideration. The central telephone number or other numbers where bomb threats have been received may be connected to tape recording apparatus. This practice is legal in nearly all intrastate circumstances since one of the two parties in the communication is aware of the recording being made. The intent is to preserve all pertinent information involved in the threat. However, it is not entirely clear that a voice print obtained from the recording would be admissible in a Court of law to identify a caller.

In certain circumstances and for certain telephone service installations it is possible to have "trap circuits" connected to telephone numbers receiving frequent bomb threats. This technique is similar to "tracing" the origin of a given telephone call. Specific details of facility responsibility in regard to implementing this technique are beyond the scope of this book.

(c) Once an area is searched and found clean, it is closed to further access.
(d) Medical personnel and an ambulance are positioned nearby in case the search activities result in detonation of the bomb.
(e) Doors and windows are left open, if possible, as this may tend to vent high pressures from a bomb blast and reduce injury from flying glass and other debris.
(f) Any suspicious items are reported to the BDU and no attempt is made to cover, disarm, or touch the suspected item.
(g) The use of radio transmitters and other devices radiating even small amounts of radio frequency energy may have to be curtailed in certain areas until the situation is resolved.

ALARM 25: GUARD STRIKE

Initial Reaction

The initial reaction is made by management based on the status of the plant. The vulnerability of the plant and the type of strike must be considered.

Since guards would be protected by the National Labor Relations Act (NLRA), the licensee must provide plant security without infringing upon the federally protected rights of the guard employee. However, many states have statutes that prohibit certain classes of employees, such as police, firemen and teachers from striking. Guards may be within that group prohibited from striking.

Assessment and Determination

To determine the threat represented by a guard strike, the reactor management requires as much information as possible about the strike. The first threat analysis here deals with an impending guard strike. The second analysis deals with a sudden, unpredictable "wildcat" strike of guards which leaves the plant without guards. Slowdown activity is dealt with in the third analysis.

The threat level determination for this type of activity involves determining where the striking guards are located, how many are involved, and if the activity involves only passive refusal to protect the plant or includes some propensity for violence.

Impending Guard Strike. The reactor management must determine what positions or rank of security officer will honor the strike. Guards holding supervisory positions are evaluated to determine if they will honor or engage in the strike.

The roster of security trained utilities operators is reviewed to determine how many, if any, employees could be utilized as temporary guards on a volunteer or other basis.

Wild Cat Strike. A wildcat strike is a labor action that is unexpected or unpredicted. However, management can usually count on knowing the possibility of a strike through employee discussions and actions.

Since Section 8(d) of the National Labor Relations Act (NLRA) requires* a notice of sixty days when the bargaining unit wishes to modify or terminate the contract, a wildcat strike is a labor action in violation of the notice. Section 8(d) of NLRA specifies that an employee engaged in a wildcat strike loses his status as an employee under the NLRA and consequently any protection under the NLRA. Therefore, the wildcat striking guards can be replaced immediately by supervisory guard personnel or other guards. In addition, the wildcat strikers may be ejected from the plant if necessary.

Labor Dispute. The problem would be not with wildcat strikers but with slowdown activity, which is difficult to detect. A strike of this type can take the form of a work slowdown or a total work stoppage. Dealing with a slowdown activity should be left to the discretion of the security supervisor.

Tactical Response

A guard tactical response as described for the other alarms or events is not relevant. Management policy on reactor security versus power needs must be made in order to consider shutting down

*There must be in effect a collective bargaining agreement in an industry which affects commerce.

the reactor and purchasing power elsewhere. Shutting down the reactor and locking up certain areas, then patrolling the protected and vital areas of the reactor with plant personnel while local police patrol the exterior must be considered.

Management or armed response personnel may be required to perform routine security functions such as patrol and alarm monitoring until the situation is resolved.

ALARM 26: KIDNAP – THREATENED OR ACTUAL

Initial Reaction

Initial reaction is made by the security supervisor and is based on the nature of the hostage, the hostage's knowledge, and the likelihood that the kidnapper can cause a significant act of sabotage or theft. In all but a few instances, the element of time works against the kidnapper and for LLEA; therefore, the security supervisor needs to determine whether immediate action is necessary. According to literature developed by the law enforcement community, attempts to alleviate the situation through SWAT actions or inflammatory conversation will, by and large, be counterproductive.

Assessment and Determination

To determine the threat, the security supervisor needs to consider the following:

The Kidnapper. The kidnapper will be one of three general types of individual: criminal (usually "caught-in-the-act"), psychotic, or politically motivated. Only the politically motivated (terrorist) kidnapper automatically represents a threat of substantial magnitude. Generally, the criminal is rational and has no desire to commit the threatened act. The psychotic is seeking gratification and can probably be neutralized by proper psychiatric actions by LLEA. However, the politically motivated kidnapper believes his act to be morally acceptable and will generally have a well-planned operation and the necessary knowledge to commit the threatened act.

The Hostage. It is important to know whether the hostage possesses the knowledge to seriously compromise the facility. However, this is not a signal for immediate action unless other factors indicate it is necessary.

The Capability of the Kidnapper. The criminal, as noted above, should not be pressed. The psychotic (unless an "insider") will not generally have the knowledge to compromise the facility. The politically motivated kidnapper must be treated as having the knowledge to commit the act. In addition, if the kidnapper (of any kind) has a knowledgeable hostage it must be assumed that the threatened act is possible.

Pertinent Data. Data pertinent to the situation includes the following:

1. number of hostages,
2. apparent type of kidnapper,
3. knowledge of hostage,
4. objective of kidnapper,
5. threatened act,
6. location of hold-up, and
7. feasibility of threatened act.

Determination of Desirability of Overt Action. Overt action will only be justified if it is determined that the threat will be carried out as quickly as possible by a politically motivated terrorist. The objective of the terrorist must be the threatened act. Where negotiation is possible it must be relied upon. LLEA is the proper authority to negotiate, and in any event LLEA is notified immediately.

Tactical Response

A guard mobilization and tactical response as described for other alarms is not relevant.

ALARM 27: SECURITY RADIO NETWORK JAMMING

Initial Reaction

The guard noticing the problem immediately telephones the CAS to determine if a security contingency is taking place and to receive orders.

Assessment and Determination

A preplanned set of procedures determines the threat level and are followed if the security radio network is jammed or becomes inoperable for any reason.

An assessment is made of the cause of the jamming or failure of the radio system. Causes of failure may be attributable to power failures, electronic failure, or the result of intentional damage. The failure may occur at the CAS control panel, at the repeater transmitter, or at the transmitter antenna. Jamming occurs when the transmit key is held down on a transmitter unit. This occurs accidentally when patrol officers place the portable transmitter in such a way that the transmit key is depressed. Cases are known where a portable hand held unit is placed on the seat of a patrol vehicle and then sat upon by a patrol officer in such a way that the key is depressed. The assessment tries to locate and check all portable and vehicle radio equipment.

Tactical Response

Regardless of the cause of the jamming or failure, the following actions are taken:

1. Many security radio networks are equipped to operate on either of two channels. All members of the security force and the radio operator in the CAS switch to the alternate channel.

An alternate radio communication system such as that used by plant maintenance personnel may be used by the security organization on a temporary basis. The use of an inexpensive citizens band system as a backup may be worthy of consideration. Word of mouth, messengers, and written messages are also options. If needed, the use of a plant public address system, warning horns, or plant lighting are used for communication of information to guards and security force personnel.

Throughout this type of threat, the security force personnel should keep the transmission of information over telephone or an alternate radio system as brief as possible. Serious consideration should be given to the use of a special set of codes distinctly different from that used up until this threat. An intentional jamming caused by adversaries indicates some knowledge of the facility radio system, including perhaps knowledge of the normal radio codes.

2. Radio repair service should be on twenty-four hour call to begin repairs if needed on short notice and is contacted if the jam

cannot be cleared within five minutes.

3. All fixed and roving patrols report to the CAS by telephone to give their position and status and to receive further instructions. Routine security patrols are maintained with guards on the alert, reporting any unusual incidents.

If it is determined that the jamming or failure is the result of intentional damage, the security supervisor initiates *Response 4* (Defend the vital area). Response team members are alerted to the possibility of either an internal or an external intruder attack.

ALARM 28: CIVIL DISTURBANCE

Initial Reaction

Plant management may have an advance warning of the time and type of any planned civil disturbance. The security force is then prepared.

However, if the disturbance is abrupt it will most likely take the form of *Alarm 5* (Someone coming over fence) or *Alarm 14* (Someone coming toward or near fence) although others, e.g. sound of explosion (#8), barrier breached (#18), jamming (#27), are also likely. Initial reaction will parallel that for the similar alarm.

Assessment and Determination

If warranted, the assessment and determination occur simultaneously with guard mobilization and tactical response. The security supervisor must assess all available information, including that outlined in *Alarm 26* (Kidnap – threatened or actual), and determine the level of threat.

Tactical Response

The actual response chosen depends upon the specific actions of the individuals involved in the disturbance. Details are given under alarms corresponding to those actions. In all circumstances, though, *Response 4* (Defend the vital area) is in effect.

Figure 27. Supervisor Checking Fence. Courtesy of Allied-General Nuclear Services.

ALARM 29: FENCE MOTION DETECTOR ALARM

Initial Reaction

View the area with CCTV since the ratio of false alarms to real alarms is often very high and this technique is faster than sending a guard to the fence area. However, sending a guard is a more reliable technique and should be done as a follow-up procedure.

Assessment and Determination

The assessment for this alarm consists of determining if the alarm is caused as a result of an intrusion or by some natural event such as snow or wind. To make this determination one guard views the area with CCTV while another guard is enroute to inspect the area of the alarm. After arriving at the scene, the guard inspects the area for signs of intrusion and inspects the area for artifacts such as snow or birds which could explain the alarm.

If intruders are discovered, the visual assessment is relayed to the security supervisor who will initiate a tactical response to the intrusion.

Tactical Response

The request for tactical response to this alarm is the result of a threat analysis indicating a need for a response. The threat analysis gives information as to what the responding guards can expect to find.

The security supervisor initiates *Response 7* (Defend the vital area with maximum confrontation). This response assures that the vital area, which is the main item of interest, is protected while the intruders are intercepted and neutralized. In addition, the protection of the vital area helps to assure that undetected intruders will not be able to assault the vital area undetected.

The information obtained from the threat analysis is used to determine where the maximum confrontation could occur, and the guards responding to the vital area portals do so via a protected route,* even if it is longer.

The control room personnel and guards unable to respond are alerted to the fact that a tactical response is underway.

Depending upon the analysis of the threat and availability of guards, a second choice of guard mobilization is *Response 5* (Defend the vital area while single guard confronts with reserve).

ALARM 30: RADIATION ALARM AT PORTAL

Initial Reaction

The initial reaction is to provide a guard who, from a safe distance, views the corrective action taken by the guard or person responsible for the portal.

---

*A protected route is through a building or along a path which shields the guards from view or from small arms fire from the intruder.

Assessment and Determination

This alarm represents an unusual circumstance that does not require an immediate response by the guard force. However, a member of the guard force should be present to evaluate any safeguard concern of this alarm.

Tactical Response

A radiation alarm at a portal requires a tactical response only after a threat analysis has determined that a response was necessary.

Since this alarm occurs at a portal where the mobility of the person causing the alarm is limited, the security supervisor implements *Response 2* (Single Confrontation with a Reserve) since there is probably a health physics monitor or a guard already nearby who could assist.

ALARM 31: FIRE ALARM

Initial Reaction

Since a security guard's sole interest is in providing security at the plant, he does not engage in fire fighting. The initial reaction is to dispatch a guard to the fire scene for the sole purpose of observing and detecting activity that might be part of a sabotage or theft of SNM attempt. The other guards throughout the plant are alerted to the existence of the fire and are placed on alert for suspicious activity by persons using the fire as a diversion.

Assessment and Determination

The assessment involves an inquiry into the cause of the fire to determine if it is related to sabotage of the plant. A second objective is to determine if the fire is part of a larger sabotage action underway.

Tactical Response

Ordinarily guards cannot dilute their safeguard activities by fighting fires. However, guards need to be on the scene to observe and detect unauthorized individuals, to escort and direct personnel, and to protect vital areas. Since fire, like the explosion of *Alarm 8*, could be a diversion, the guard must be especially alert to detect unauthorized activity elsewhere during the fire.

No response is required by the guard force unless the fire is found to be or is suspected to be part of a larger sabotage or theft effort. The security supervisor then elects *Response 7* (Defend the vital area with maximum confrontation). Some guards normally designated to confront are used to patrol and observe other areas of the plant for suspicious activity while the rest of the confrontation guards assist with security matters around the fire.

ALARM 32: LIFE–THREATENING–ACCIDENT ALARM

Initial Reaction

A guard is dispatched to the scene with the express purpose of determining if there are any security matters connected with the accident. All other guards are alerted to the possibility of sabotage or theft action being attempted or taking place during a diversion caused by the accident.

Assessment and Determination

The assessment for this alarm requires the security supervisor to evaluate the possibility that the accident may be a diversion while sabotage or theft is being committed elsewhere or that the accident is somehow part of an overt attempt to sabotage or steal SNM.

Tactical Response

The tactical response for this alarm would be the same as that for *Alarm 8* (sound of explosion) or *Alarm 31* (fire). The primary concern is for the guard force to be aware that the accident may be a diversion.

Chapter 4

FACILITY EMPLOYEE RESPONSE TO AN INTRUSION ALARM

OBJECTIVE

TO maximize a reactor facility's response to intrusion the nonsecurity facility personnel could be asked to volunteer either as "tactical assistance personnel" or as volunteer "assistance personnel" to help the guard force.

Assistance personnel would attend to nonsecurity matters such as securing facility equipment and directing the activities of other personnel. In addition, the assistance personnel would lock vital areas, attend to the operation of the reactor, direct nonessential personnel to safe quarters, and attend to the facility operation problems caused by the intrusion. The facility operators required to operate the plant would be classified as assistance personnel but would not be subject to the volunteer process.

The tactical assistance personnel can act as observers for the guard force, carry messages, provide technical advice at the telephone switchboard and perform other tasks requested by the guard supervisor.

Employees not vital to the safe operation of the plant or involved in tactical or assistance activities would be required to go to a place of safety where they would be less vulnerable to being taken as a hostage.

REGULATORY REQUIREMENTS AND LEGAL RESTRAINTS

Regulation

No specific regulation requires, prohibits or allows nonguard employees to take actions that would minimize the results of an armed intrusion. The Code of Federal Regulations (10CFR 73.55 (h) (2))specifically requires that personnel acting as guards or armed response personnel be armed and trained. In addition, 10CFR 73.55 (b) (4) states that "the licensee shall not permit an individual to act as a guard, watchman or armed response individual unless such individual has been properly trained and qualified and has demonstrated: (1) an understanding of the licensee's security procedures, and (2) the ability to execute all duties required of him by such procedures." However, the kind of activity recommended here for an employee is similar to that traditionally expected from volunteer employee-fire-fighters, first aid teams, or disaster teams and is not the same as that required of a guard.

Legal Problems

An employee would not be asked to perform guard functions and would not be exposed to the same liabilities that are associated with guard work. However, both the guard and the employee are private citizens and agents of the employer. As such, they would both be subject to the same liabilities except for one principle difference. If a lawsuit for negligence should arise, the employee would be held to (or expected to act according to) a standard expected of a reasonably prudent man who was similarly trained. To rephrase the standard, a guard would be expected to act in a manner expected from a reasonably prudent, similarly trained guard, and an operator would be expected to act in a manner expected from a reasonably prudent similarly trained operator.

Although it is not intended that employees act as guards the following discussion on the legal problems affecting guard response to intrusions is given to illustrate problems that an employee could expect if he were required or were asked to act in a capacity of a guard.

There are general legal rules which can be briefly given here as a guideline to shape the action of an employee requested to assist a guard. However, each state has its own individual rules that may vary the general legal rule, causing the employee action to perhaps be different for that state.

The principal legal concerns affecting the employee are those regarding (a) the use of deadly force, (b) corporate liability for actions of the employee, (c) the arrest powers of the employee, (d) the criminal and civil assault and battery, and (e) trespass.

The Use of Deadly Force

Deadly force may be defined as (a) force which its user uses with the intent to cause death or serious bodily injury to another, or (b) force that he knows creates a substantial risk of death or serious bodily injury to another.

There are general legal theories that may be used to justify the use of deadly force and prevent criminal liability by the guard and his employer:

- self-defense
- defense of another
- defense of property
- law enforcement
  - effecting a lawful arrest
  - preventing an escape from custody
  - crime prevention and termination of crime
- necessity

Because of the importance of the subject of the use of deadly force, the following must be considered:

(1) Self-Defense:

Briefly stated, an employee as a private citizen may defend himself against deadly force if deadly force is directed at his person.

If the employee injects himself into a gun battle to rescue other guards or employees, he may in some states be forced to accept the same legal status of the guard being rescued and if the guard being rescued did not have a right to use deadly force then the rescuer would not have a right to use deadly force. Some states allow the rescuer to make a reasonable mistake as to the status of the guard being defended.

In addition, there is some disagreement between states regarding the necessity of retreating if it is safe to do so, before using deadly force in self-defense.

(2) Defense of Another:

Some states require that the defense of others be limited to defending those with a special relationship to the defender (such as husband, wife, parent, and child). This defense would not be available to an employee arguing that he was defending the whole city against nuclear threat, because he would not have the requisite relationship to those defended. However, this same argument might be available under another theory of justification called necessity.

(3) Defense of Property:

The use of deadly force is never reasonable in defense of property except where the unlawful interference with property is accompanied by threat of deadly force. If deadly force is threatened, then the issue is not defense of property but it is self-defense.

Under the general rule, an employee could use deadly force against a trespasser who was committing a dangerous felony (dangerous felonies are limited to robbery, rape, mayhem, burglary, arson, and kidnapping).

(4) Law Enforcement:

An employee who is not a police officer is justified in using reasonable force to prevent or terminate what he reasonably believes to be the commission of a misdemeanor amounting to a breach of the peace or of a felony. It is not reasonable for a private person such as a guard or an employee to use deadly force to prevent the commission of a crime unless the crime is a dangerous felony.

A private guard or employee acts at his peril in using deadly force. There is no privilege to use deadly force, no allowance for a reasonable mistake (such as is allowed for peace officers), if it turns out that the person against whom the deadly force was used did not commit a dangerous felony.

(5) Necessity:

The defense of "necessity" is a judicial and societal policy rule that the lesser of two evils is preferable if a choice must be made in an emergency. As a matter of public policy, the law would promote the higher values at the expense of the lesser values.

This argument of the lesser of two evils might be used to justify deadly force when it is used to save a large number of people from the consequences of sabotage of a reactor.

Necessity is an accepted defense in some areas of the law and should be acceptable as a strong defense for the use of deadly force by an employee. However, a practical problem arises in the use of this argument because the employee would have to show the imminence of a disaster to prove necessity.

The failure of an employee to be able to qualify for any of these defenses for the use of deadly force would mean that the employee could be charged and prosecuted either for murder, manslaughter, criminally negligent homicide, or assault with a deadly weapon.

Corporate Liability for Actions of the Employee

It is now generally conceded that a corporation/utility may be held criminally liable for conduct performed by an agent (a guard or employee) of the corporation/utility acting in its behalf within the scope of his employment. The Supreme Court decided in *New York Cent. & H.R.R. v. United States* 212 U.S. 481, 29 S. Ct 304, 53 L. Ed. 613 (1909)* that constitutionally "the act of the agent, while exercising the authority, delegated to him—may be controlled, in the interest of public policy, by imputing his act to his employer and imposing penalties upon the corporation for which he is acting."

Many crimes for which the punishment is death or imprisonment are statutorily constructed so as to permit fines on corporations.

It has been established that a corporation may be guilty of manslaughter *(State v. LeHigh Valley Ry Co.,* 90 NLJ.L. 372, 103 A 685 (1917) and there "is no logical reason why a corporation should not equally be able to incur criminal liability for murder," Mueller, Means Rea and the Corporation, 19. Pitt L. Rev. 21, 23 (1951).

Despite these comments the corporation/utility criminal liability for acts of the employee are not entirely clear. It is clear, however, that a distinct possibility of liability exists.

---

*This case despite its age is still good law

Arrest Powers of the Employee

The typical arrest statutes allows either a police officer or a private person (such as a guard or employee) to make an arrest without an arrest-warrant, if in his presence a felony or breach of the peace is committed. A police officer may make an arrest without an arrest warrant, for a felony not committed in his presence if he has reasonable grounds to believe (a) that a felony has been committed, and (b) that the person to be arrested committed the felony. For a felony arrest, a private person, however, is given no margin of error, he is permitted to arrest without a warrant only if the felony has, in fact, been committed. Neither a police officer or a private person may arrest for misdemeanors committed outside their presence. In general, when a private person such as a guard or employee makes an arrest on his own, he acts at his peril.

As seen above, the arrest powers of the private guard are no different from those of any other private person such as an employee. However, because the guard usually knows what these powers are and the individual being arrested thinks the private-person-guard has the same arrest powers of a police officer, there is usually no challenge to the guard's arrest technique. However, an employee could not expect the same degree of cooperation should he attempt an arrest.

Assault and Battery

Assault and battery are two different offenses that are codified in most states.

Criminal battery is briefly defined as the unlawful application of force to the person of another. Criminal battery is usually a misdemeanor. There is also a civil battery which is defined as the intentional, unconsented to offensive touching of another.

Many states have statutes that define aggravated criminal batteries as felonies. An aggravated criminal battery would be an assault with a dangerous weapon or an assault that creates grievous bodily harm.

An employee would have to concern himself with the possibility of personally being liable for a criminal battery or civil battery if an innocent trespasser was roughly treated or unnecessarily handcuffed and offensively touched, as in a search.

Battery involves touching a person, assault is more complicated in that it involves an attempted battery or the intentional creation of an apprehension of a battery. Simply put, an assault may be an intentional attempt to touch or it may be the intent to frighten the victim causing him to have a reasonable apprehension of immediate bodily harm. Thus, if an employee would suddenly threaten an innocent intruder with a revolver saying, "Halt or I will shoot," there may be a criminal or civil assault charge which the intruder could bring against the employee, who is a private person.

Trespass

The trespasser is defined as "a person who enters or remains upon land in the possession of another without a privilege to do so, created by the possessors consent or otherwise."[10]

The general rules for trespassers involves the requirement that the land occupier must use reasonable care for the trespasser's safety, once his presence is known. This is lineally descended from an older rule, where the possessor of land was not free to inflict unreasonable intentional injury upon his unwelcome visitor. A trespasser, while he may be a wrongdoer, is not an outlaw,* and an intentional, unprivileged battery upon him was too much to be tolerated even by the great veneration of the English courts for rights in land. The land possessor was not permitted to set traps for the trespasser or to use unreasonable force to expel him from the premises. Nor, in later cases, was he allowed to injure him negligently by an act specifically directed toward him, or recklessly by conduct in conscious disregard of his peril. The usual phrase which summarized the exception was that there was liability for any "willful or wanton" injury to a trespasser.

The great majority of courts have discarded "willful or wanton" entirely as a limitation, and have said outright that once the presence of the trespasser is discovered, there is a duty to use ordinary care to avoid injuring him, as in the case of any other human being. The land occupier is then required to govern his active conduct, such as running a train, conducting a circus, or operating an elevator, with the caution of a reasonable man for the trespasser's safety.

---

*In most states trespass in only a misdemeanor.

The general discussion above on trespass means that the utility guard force and employees have a duty to act in a reasonable manner in determining the level of threat and in responding to an intrusion. Requiring an untrained employee to act in a reasonable manner in the face of possible sabotage to the reactor plant is a difficult requirement.

EMPLOYEE ACTION REQUIRED FOR EACH ALARM

Introduction

The activity which employees can be expected to accomplish during an intrusion or alarm depends on the kind of expertise the employee has and on what kinds of things need to be protected.

a. As shown in the Organization Chart shown in Appendix A, the types of employees range from the plant superintendent to technicians.

b. The things that need to be protected fall into the three classes; things, areas and people.

(1) Things include specific pieces of equipment, which can be locked, as well as other items such as computer monitors, which are necessary for the safety of the plant.

(2) Areas consist of places like vital and protected areas which can be closed or locked.

(3) People consist of individuals which could be used as hostages or as sources of information to aid in sabotage.

Objective

Although the employee does not have as a principal employment responsibility the obligation to interpose himself between the intruder and plant, he does have the capability of doing so, or impeding the intruders progress by activating mechanical and electrical impediments built into the plant.

The employee has the capability of contributing to the delaying of the intruder and reducing the vulnerability of the plant, while the guards are delaying the intruder by the use of deadly force or other means.

The activities that an employee could perform:

a. Close doors.

b. Lock doors.
c. Activate alarms out of service because of work being performed nearby.
d. Withdraw from and lock vital areas.
e. Act as observer for the Central Alarm Station as requested.
f. Assist guards as requested by guard supervisor.
g. Begin plant shutdown.
h. Man special stations such as first aid, fire department, and radiation monitoring.
i. Begin special monitoring of vital reactor parameters.
j. Man the secondary control room.
k. Assist other employees to safety.

Recommendations

Discussion

The licensed power reactor facilities' measures to thwart sabotage and theft attempts should include measures that assure that all personnel on-site have a prearranged response to intrusions. Regardless of the exact nature of an individual response, e.g. hiding, retreating, checking equipment, etc. a planned response will be preferable in every respect to an extemporaneous response. Therefore, the response procedures suggested here include not only counter-intrusion measures, but also include "self-defensive" measures designed to mitigate possible harm to employees.

In order to properly design such procedures, the guard force duties must be understood so that employee can complement, not distract, the defensive actions by the guard force. Employees should "assist" the guard force, and take defensive actions only where absolutely necessary. The guard force is "duty bound" to protect the reactor from sabotage because that is the primary job responsibility of the guard as an employee. This is not true of employees whose primary job objective is something other than prevention of sabotage or theft.

However, measures may be taken by the employees, which while not "guard" oriented, can help thwart the aims of an adversary. These involve mainly securing entrances, exits, providing communications and observations, as may be desirable according to specific site needs.

The scope of these recommendations is confined to development of procedures for employees in the event of a security contingency. A security contingency does not include emergencies such as fire, flood, strike, riot or civil disorder. It is important to note that employee response procedures should be the same in every case and not graded. Having a single response makes the procedure less difficult to remember and check during drills.

To properly develop a nonguard response employee positions and actions are classified into three categories. This scheme of classification will describe methods of designating personnel, their duties, and actions that can be taken.

In addition, time plays a very important factor. For example, an employee response will be far more effective if adequate time exists for a thorough response. It is assumed here that plant shutdown generally used in safety contingencies can, with minor modification, be employed as a measure. It will be assumed that employee response described in this book will be a short-term response begun within only a few minutes or immediately after the signaled alarm.

The analysis in this book depends on three categories of employees: (1) "tactical assist personnel," (2) "assist personnel," and (3) "other personnel." The category of the employee depends on whether an individual employee (1) volunteers to assist the defensive response, or (2) volunteers to assist the plant or other personnel or to retreat to a position of safety or (3) simply has been designated to retire to a position of safety. The "tactical assist personnel" duties will be to assist the guard force in nonsecurity matters as directed by the security personnel. The "assist personnel" have duties exclusive of the guard responsibilities such as traffic control or escort. The "other personnel" must retreat to a safe position as directed by the "assist personnel." A scheme for assignment of duties is as follows.

Duties

Because equipment is the principal target of a safeguards contingency event, it must, as a primary objective, be secured. This activity would be more properly and efficiently done by employees whose primary duties are operators than by members of the guard force. Operators, technicians, or other employees who routinely deal with operations are best suited to this task.

The assignment of personnel to specific actions will require designating employees as either "tactical assist personnel," "assist personnel" or "other personnel" defined as follows:

- *Tactical Assistance Personnel* – those personnel who are assigned specific duties assisting the overall guard effort to thwart the objectives of adversaries.
- *Assist Personnel* – all personnel who are instructed to assist others to retreat to positions of safety as per the contingency plan and secure important equipment as needed.
- *Other Personnel* – personnel not in the above two classes who must retreat.*

Tactical Assistance Personnel. The process of selecting personnel to fill tactical assistance roles is one that should involve the same "volunteer" circumstances as a typical site fire fighting team. Security, as safety, is a compelling responsibility of all employees and will be viewed as such by many employees. However, it must be emphasized that with the exception of health/physics personnel all "tactical assistance personnel" should be selected at their option. If volunteers are not available to perform these tasks, the licensee may encourage participation by making changes in job descriptions and or allowing increased compensation to allow for tactical assistance personnel involvement. Compulsory participation in contingency response may result in problems in employee relations, union negotiation, and may raise legal issues, all of which can be mitigated by a volunteer program.

Employees whose routine duties include plant operation should be expected to participate in contingency plans as tactical assistance personnel. In addition, health/physics personnel should be expected to participate in the interests of health and safety. The majority of the health/physics assistance duties may be upon reentry to a compromised facility, however, some duties will require response during a contingency event.

Tactical assistance personnel have diverse duties. Two basic

---

*The term "retreat" is not used here to connote uncontrolled fleeing but rather assumption of a position of safety, within the facility.

principles in designating tactical assistance personnel duties are (1) to free a security guard from a nonjeopardy duty so that more defensive strength can be exerted by the guard force, and (2) to provide information, communications and general assistance as requested by the security supervisor.

Since the fundamental response tactic of any guard force at a nuclear facility is to deter the adversaries until LLEA can arrive, tactical assistance personnel may be able to assist in maintaining communication with LLEA when the guard is needed elsewhere. In addition to being able to free a guard, or at least allow him possible freedom of movement, an employee supervisor knowledgeable of the facility areas may be of great value at the central communications point. Tactical assistance personnel may also be able to go to rooftop positions with high outward visibility and safely report on movements of adversaries. This would be a particularly valuable technique when a portion of the protected area is not subject to attack, but still must be kept under surveillance in case of a diversion ruse. The observation and communication locations must be selected on the basis of superior defensive positioning and have an immediate escape route available nearby. Individuals selected as observation and communication tactical assistance personnel should be trained in the use of communications equipment, e.g. hand-held transceiver, language and codes and surveillance equipment, e.g. binoculars, view scope. Since possible scenarios are varied, the deployment of observation and communications personnel must be directed by the Central Alarm Station at the time of intrusion. Therefore, a central meeting point (in a position of safety) may be required for observation and communication tactical assistance personnel. In addition, an expedient method of communicating assignments must be employed.

Assistance Personnel. Duties for employees designated as assistance personnel should be developed along the principle that the assistance personnel response should minimize the potential for injury to plant personnel as well as protect the plant from danger.

Certain assistance personnel may be used in directing flows of (nonassistance) other personnel to positions of safety. Such retreat routes must be well known to the individuals directing personnel traffic since they may have to exercise a certain amount of judgment in choosing a route. For the sake of safety as few a number of routes as possible should be considered, and judgment should be based on information provided by the Central Alarm Station describing the location and nature of the intrusion.

Another assistance-personnel duty could be the locking of gates and doors where no immediate significant probability of jeopardy exists. This means that certain assistance personnel could have primary responsibility for securing all entrances and exits that are not in the general area of the intrusion. Short instructions on locking and unlocking gates and doors would be sufficient to insure expeditious securing of openings.

The concept of First Aid assistance personnel duty probably already occurs in the fire and safety contingency plans. It should be reasonable to transfer first aid plans directly into the employee response plan. The employees required to continue the safe operation of the plant would also fall into this class.

Other Personnel. It is clear that some on-site employees cannot, under most foreseeable circumstances, assist in any way. Many employees will be more properly instructed to retreat to a position of safety.

All personnel not involved in the two above classes would be required to take shelter in some safe area to avoid being taken as hostages.

Post-intrusion Plan

After an intrusion, a state of disorder may exist. If an evacuation of the facility took place, a reentry is necessary. It is important to stress that those in a position of safety should remain in their safe position until instructed by an authorized individual to reenter the plant. The logistics of possible intrusion scenarios are so numerous that the state of defense should be maintained until an "all-clear" signal is given. Assistance personnel should be given personnel traffic responsibility for instructing personnel away from positions of safety after the all clear signal has been given.

Post-intrusion Reentry. It would be desirable to have employees involved in the post-intrusion reentry problems quickly help ascertain whether bombs or incendiary devices have been left in the reactor. An employee search team could do a great deal to shorten the period of time necessary to assure that the facility is secure.

Where reentry involves the use of outside repair personnel, consultants, or others not employed by the facility, employees could serve as escorts. For instance, law enforcement officials would need access to the facility generally to accumulate evidence, e.g. fingerprints, etc. Since law enforcement personnel lack the appropriate clearance, escorts would be required.

Health/physics personnel should be first to reenter in order to monitor any ambient releases.* It should be borne in mind by the reentry team that a latent possibility exists for employee injury, e.g. bomb, or presence of an adversary in hiding.

A reentry team should be composed of assistance personnel trained for reentry according to specific site requirements and alerted to the particulars of the intrusion.

Summary

Employees should be made aware of the contents of the contingency plan. It should be made clear that participation is optional (voluntary) and that all employee duties will be supportive and, to as great a degree as possible, noncombative in nature. The need for personnel control and planned assistance responses are essential to the safety of employees.

All employees under consideration as assistance personnel should then be brought together to describe the nature of contingency planning and aspects involving employees. A breakdown of responsibilities should be given. A sample of duties is as follows:

- Tactical Assistance Personnel
  Observers
  Messengers
  Switchboard Operators
  Control of Prisoners
- Assistance Personnel
  Personnel Traffic Control
  Central Communications to other employees

*This duty would be similar to the duty health/physics has after a fire.

Gate and door security
First Aid
Reentry Team
Bomb Squad
Escorts

Participants should be given a chance to indicate preferences, however, the security supervisor and facility administration must make a final determination of suitability. Criteria for determining suitability should be developed in advance of assigning employee positions.

The existence of an employee training program is implicit in constructing an employee response.

Specific Procedures by Organization

When an intrusion alarm is given by the CAS, the employees should begin an orderly preset procedure to diminish the susceptibility of the plant to sabotage or theft. Since the employees and probably the guard force have no way to determine the intended specific objective of the intruder, the employee procedure should be the same for every intrusion alarm.

In addition, since the time from the intrusion alarm and the appearance of the intruder could be as little as 2 or 3 minutes, the employee must begin the intrusion procedure immediately upon hearing the alarm.

The specific procedures by organization would be:

Reactor operations personnel

(1) Assure that staff adequate to cover an emergency is in the control room
(2) Dispatch an operator to the secondary control room
(3) Dispatch a junior operator to verify vital areas are closed and locked
(4) Request that any work in vital areas be temporarily halted and that personnel withdraw from and lock vital areas
(5) Lock the control room door
(6) Activate all remote locks, after determining it is safe to do so
(7) Confer with plant management about advisability of

shutting down the plant

(8) Begin special monitoring of parameters selected as indicators of possible sabotage or theft.

(9) Be prepared to shut down the plant immediately, if required.

Plant Management Personnel

(1) Obtain status report from security supervisor and consider advisability of plant shutdown.

Technical Support. QA, Engineering Dept., Maintenance, etc.

(1) Man special emergency stations such as first aid, fire, radiation monitoring.

(2) Make available several personnel who can assist the guards as requested in certain tasks such as observers or message carriers.

Administrative support

(1) Lock files of drawings and other containers of sensitive materials

(2) Lock offices to protect personnel from being made hostages.

(3) Withdraw to a position of safety.

Special Procedures for Tactical Assistance and Assistance Personnel

Tactical Assistance Personnel

(1) Stop work or prepare current work for stopping.

(2) One supervisory individual assigned to direct tactical assistance personnel should direct efforts to provide assistance to guard effort.

(3) In the event that the intruders have successfully penetrated the facility and are closing on the vital areas, the tactical assistance supervisor (always in contact with CAS) should call the control room for advice and direction in efforts, effect locking of vital areas and direct other tactical assistance personnel to prearranged positions.

(4) The supervisor should direct nonessential assistance personnel to retreat to a prearranged position of safety.

(5) Observation and communication (O&C)

(a) Those assigned to Observation and Communication (O&C) should gather at a central collection point (inside the facility) where the CAS will designate areas where O&C are appropriate.

(b) O&C personnel should report to the prearranged O&C positions and maintain continued vigilance on the area in question.

(c) If any suspicious activity begins to occur, it should be reported to CAS immediately. It is important to note that the timeliness of the warnings from O&C personnel weigh heavily in determining the defensive strategy chosen by the guard force.

(d) If appropriate O&C personnel should retreat from their positions via prearranged routes to a position of safety.

(6) Gate and Door Security (G&D)

(a) G&D personnel should be instructed by the CAS which areas should be closed off. The determination by CAS should be made on the basis of the safety of G&D personnel.* Safety contingencies should conform, wherever reasonable, to exisiting plans.

(b) First Aid personnel may be required to report to areas of fighting. If so, the CAS should direct movement of First Aid personnel so as to minimize risk of injury to responding parties.

Assistance Personnel

(1) Personnel traffic control

(a) Those directing personnel traffic should immediately establish contact with the CAS to determine which of the prearranged routes to a position of safety are appropriate, based on an assessment from CAS.

(b) All other personnel should report to a collection point (within the facility) for instructions on an orderly retreat to a position of safety.

---

*While a door may offer substantial resistance, a gate which is part of the linkchain perimeter fence will not. It may be a needless sacrifice of resources to lock a gate.

(c) Those directing personnel traffic should accompany other personnel and maintain listening contact with CAS.

(d) It is imperative that other personnel stay in the prearranged position of safety until an all-clear signal is relayed from the CAS.

(2) Central Communications

(a) This duty should be performed by a supervisor familiar with the layout of the plant and process area (material access area).

(b) The individual designated for central communications should check with a personnel traffic director and, if possible, go to CAS to assist.

REFERENCES

1. U.S. Code of Federal Regulations, Title 10, Part 73.1(1). Available for purchase from Superintendent of Documents, U.S. Government Printing Office, Washington, D.C. 20402.
2. L.A. Suber, Jr., "Adversary Characterization for Security System Evaluation," SAND75-0627, prepared for ERDA by Sandia Laboratories, April 1976. Available from National Technical Information Service, Springfield, Virginia 22161. Price: Printed Copy $4.50; Microfiche $2.25.
3. A. Fainberg and A.M. Bieber, Jr., Brookhaven National Laboratory, "Barrier Penetration Database," p.24, USNRC Report NUREG/CR-0181 Rev. 1, November 1978. Available for purchase from National Technical Information Service, Springfield, Virginia 22161. Price: Printed Copy $5.25; Microfiche $3.00.
4. "Nuclear Sabotage," pp. 29-36, *Bulletin of the Atomic Scientists,* October 1976. Available in public libraries.
5. U.S. Code of Federal Regulations, Title 10, Part 100.11. Available for purchase from Superintendent of Documents, U.S. Government Printing Office, Washington, D.C. 20402.
6. U.S. Nuclear Regulatory Commission, "Draft Safeguards Contingency Planning, An Introduction," Step P8 (p. P8-3), November 1976. Available in NRC Public Document Room, Washington, D.C., for inspection and copying for a fee.
7. U.S. Nuclear Regulatory Commission, "Final Site Security Training Manual," USNRC Report NUREG-0464, Vol. 3, pp. III-45-III-54, Lesson Plan-402 Rev. 0, Step 3.C.(5), October 1978. Available for purchase from National Technical Information Service, Springfield, Virginia 22161. Price: Printed Copy $7.25; Microfiche $3.00.
8. U.S. Department of the Army, Headquarters, "Field Manual Physical Security," FM 19-30, pp. 10-6-10-9, E-1, November 1971. Available for purchase from Superintendent of Documents, U.S. Government Printing Office, Washington, D.C. 20402.
9. U.S. Nuclear Regulatory Commission, Regulatory Guide 1.1011, Rev. 1, March 1977. Requests for single copies of issued guides (which may be reproduced) or for placement on an automatic distribution list for single

copies of future guides in specific divisions should be made in writing to the U.S. Nuclear Regulatory Commission, Washington, D.C. 20555, Attention: Director, Division of Document Control.

10. Prosser, Second Restatement of Torts; Handbook of the Law of Torts (4th Ed., 1971) page 357.

BIBLIOGRAPHY

Adams, K. and Trujillo, A. *Tentative Job Analysis for a High Level Fixed-Site Nuclear Security Officer.* SAND 77-1400. Albuquerque, NM: Sandia Laboratories, 1977.

Bassiouni, M. Cherif. *Citizen's Arrest.* Springfield, IL.: Charles C Thomas, Publisher , 1977.

Blalock, Joyce. *Civil Liability of Law Enforcement Officers.* Springfield, IL.: Charles C Thomas, Publisher, 1974.

Boozer, D. and Engi, D. *Simulation of Personnel Control Systems with the Insider Safeguards Effectiveness Model (ISEM).* SAND76-0682. Albuquerque, NM: Sandia Laboratories, 1977.

Edelhertz, H. and Walsh, M. *The White Collar Challenge to Nuclear Safeguards.* NUREG-0156. Seattle, WA: Battelle Human Affairs Research Center, 1977.

El-Bassioni, A., Harris, L., Kull, L., and Lobner, P. *Protection of Nuclear Power Plants Against Sabotage by Two Insiders.* SAI-77-965-LJ. LaJolla, CA: Science Applications, Inc., 1978.

Engi, D. and Boozer, D. *The Use of ISEM in Studying the Impact of Guards Tactics on Facility Safeguards System Effectiveness.* SAND77-0410C. Albuquerque, NM: Sandia Laboratories, 1977.

Hacker, F.J., M.D., *Crusaders, Criminals, Crazies – Terror and Terrorism in Our Time.* Toronto, Canada: George J. McLeod Limited, 1976.

Huber, R.K., Jones, L.F., and Reine, E. *Military Strategy and Tactics.* New York: Plenum Press, 1974.

Hulme, B. and Holdridge, D. *SPTH3: A Subroutine for Finding Shortest Sabotage Paths.* SAND77-1060. Albuquerque, NM: Sandia Laboratories, 1977.

Hulme, B. *MINDPT: A Code for Minimizing Detection Probability Up to a Given Time Away from a Sabotage Target.* SAND 77-2039. Albuquerque, NM: Sandia Laboratories, 1977.

International Energy Associated Limited for U.S. Department of Energy. *Nuclear Safeguards Technology Handbook Dec. 1977.* Washington, D.C.: U.S. Government Printing Office, 1978.

Kaul, D. and Sachs, E. *Adversary Actions in the Nuclear Power Fuel Cycle: I. Reference Events and Their Consequences.* SAI 121-612-7803. Draft. LaJolla, CA: Science Applications, Inc., 1977.

Kull, L., Harris, L., and Lobner, P. *Protection of Nuclear Power Plants Against Sabotage by an Insider.* SAI-77-868-LJ. LaJolla, CA: Science Applications, Inc., 1978.

LaFave, W.R. and Scott, A.W., Jr.. *Criminal Law.* St. Paul, MN: West Publishing Co., 1971.

Law Enforcement Assistance Administration for U.S. Department of Justice. *Model Security Guard Training Curricula.* McLean, VA: Hallcrest Press, 1978.

Milton, C.H., et al. *Police Use of Deadly Force.* Washington, D.C.: Police Foundation, 1977.

National Advisory Committee on Criminal Justice Standards and Goals. *Private Security-Report of the Task Force on Private Security.* Washington, D.C.: U.S. Government Printing Office, 1977.

Prosser, W.L. *Law of Torts.* St. Paul, MN: West Publishing Co., 1972.

Siljander, R.P. *Fundamentals of Physical Surveillance.* Springfield, IL.: Charles C Thomas Publisher, 1977.

U.S. Atomic Energy Commission, Division of Security. *U.S. Statutes of General Interest to Security Officers in the Atomic Energy Program.* Washington, D.C.: U.S. Government Printing Office, 1970.

U.S. Marshals Service. *Security of Special Nuclear Materials – A Study for the U.S. Nuclear Regulatory Commission.* Washington, D.C.: U.S. Marshals Service, 1975.

Weaver, Leon H. *Industrial Personnel Security.* Springfield, IL.: Charles C Thomas Publisher, 1964.

APPENDIX

Typical Organizational Chart for a Nuclear Power Plant

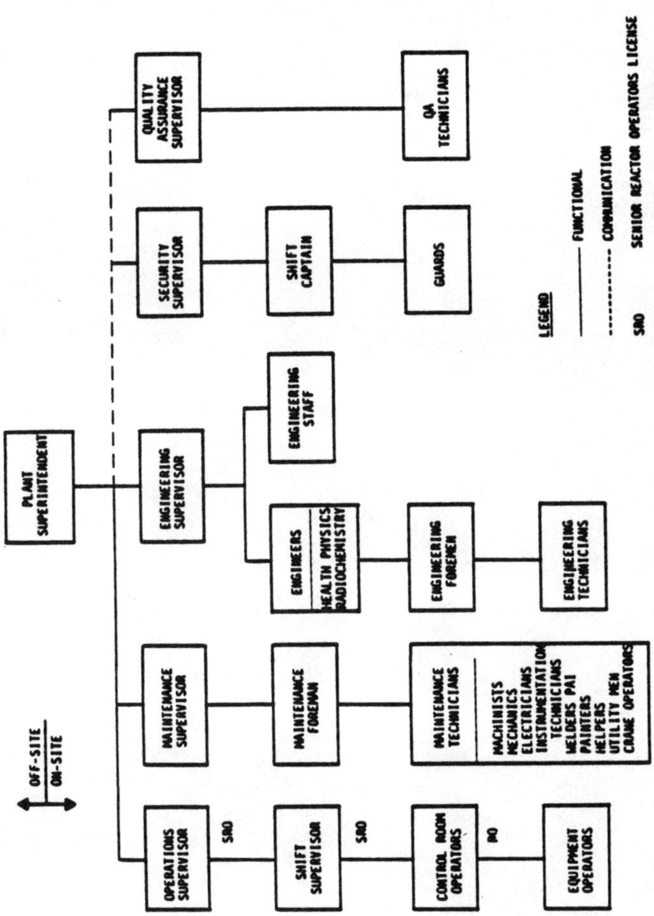

Figure 28. Typical Organization of Reactor Personnel

GLOSSARY

Adversaries: Potential saboteurs or thieves of SNM. Persons who, for whatever reasons, attempt to sabotage a nuclear facility or steal SNM.

Aids to Sabotage: Any hand carried tool, explosive, weapon or device (such as a wire cutter) that would aid in the intended sabotage task.

Barrier: Defined in 10CFR73.2(f):

(1) Fences constructed of No. 11 American wire gauge, or heavier wire fabric, topped by three strands or more of barbed wire or similar material on brackets angled outward between 30° and 45° from the vertical, with an overall height of not less than eight feet, including the barbed topping.

(2) Building walls constructed of stone, brick, cinder block, concrete, steel or comparable materials (openings in which are secured by grates, doors or covers of construction and fastening of sufficient strength such that the integrity of the wall is not lessened by any opening), or walls of similar construction not part of a building, provided with a barbed topping described in paragraph (f) (1) of this section of a height of not less than eight feet.

(3) Ceilings and floors constructed to offer resistance to penetration equivalent to that of building walls described in paragraph (f) (2) of this section.

Barrier Breach: A condition where a barrier has been successfully bypassed by an intruder.

Central Alarm Station (CAS): A continuously manned and physically protected room where all alarms enunciate and from which all tactical responses should initiate.

Civil Disturbance: A passive nonviolent action by two or more persons together that has not reached the proportions of a riot.

Contingency: A security-related incident such as an intrusion into the protected area.

Demeanor: An outward manner displayed by a person or persons.

Emergency: An event, other than a security contingency, that an intruder could use to his advantage in gaining entrance to or furthering his attempts at sabotaging a nuclear facility or stealing SNM. Events such as, but not limited to, strikes, riot, civil disorder, fire, explosion, life-threatening chemical accident, tornado, hurricane, flood and earthquake are considered as emergencies.

Guard: A uniformed individual armed with a firearm whose primary duty is the protection of special nuclear material against theft and/or the protection of a plant against radiological sabotage. See 10CFR73.2(c).

Guard-Needs-Aid Alarm: An alarm that may be covertly operated and that alerts the central alarm station when a guard is in a condition of duress.

Health and Safety of the Public: A concern for preventing a major health hazard to the public.

Initial Reaction: The activity required to place a member of the guard force into a position to determine the level of threat announced by the alarm.

Insider: An individual normally employed by, connected with, or legally admitted to the nuclear facility.

Intrusion Alarm: A tamper indicating electrical, electromechanical, electrooptical, electronic or similar device which will detect intrusion by an individual into a building, protected area, or vital area and alert guards or watchmen by means of actuated visible and audible signals.

Isolation Zone: Any area, clear of all objects which could conceal or shield an individual, adjacent to a physical barrier, which is monitored to detect the presence of individuals or vehicles within that area.

Local Law Enforcement Agency (LLEA): The police agency or agencies having jurisdiction over the locale of the reactor.

Maximum Guard Force Response: Response as if alarm announced a clear and present adversary threat.

Monitor: A device or technique used to control or record a process. To scrutinize or check systematically.

Nuclear Facility: As used in this book – Boiling Water Reactors or Pressurized Water Reactors licensed by the United States Nuclear Regulatory Commission under 10CFR Part 50 and used to generate electrical power. (Nuclear Reactor as defined by 10CFR-50.2 (k) "means an apparatus other than an atomic weapon designed or used to sustain nuclear fission in a self-supporting chain reaction") or any U.S. Department of Energy nuclear facility.

Operations Superintendent: The individual who is responsible for operating details of the nuclear facility. The operations superintendent reports to the plant manager.

Overt: An action that is open and observable, not concealed or hidden.

Physical Security Plan: Written security procedures that document the structure of the security organization and detail the duties of guards, watchmen, and other individuals responsible for security.

Plant Manager: The senior representative of management who is normally at the nuclear facility during the regular work week and has the responsibility for the entire plant and staff.

Plant Operator: Licensed individuals who operate the nuclear facility under the management and direction of the shift supervisor (also called shift engineer or watch engineer).

Portal: A doorway entrance or gate.

Position of Safety: A physical state where a guard can assess a particular threat while maintaining assurance against a reasonably suspected assault.

Position of Strength: A physical state where a guard is situated so as to maximize his defensive impact.

Protected Area: An area encompassed by physical barriers and to which access is controlled. The first immediate area inside the outermost guarded barrier around a Nuclear facility.

Radiological Sabotage: Any deliberate act directed against a plant in which a nuclear activity is conducted, or to any component of such a plant, which could directly or indirectly endanger the public health and safety by release of radiation, other than such acts by an enemy of the United States, whether foreign government or other person.

Response Plan: A preconceived and prepared list of steps to be taken to mitigate the potential security and safeguard loss during a contingency.

Search: To make a thorough examination of or to look over carefully in order to find something. To explore.

Secondary Alarm Station (SAS): The alternate location for communication of alarms as required in 10CFR73.55e(1).

Security Contingency: See Contingency.

Security Supervisor: The highest ranking member of the security organization always on duty at the nuclear facility. The security supervisor exercises immediate management control over all security guards on duty. The security supervisor normally reports directly to the plant manager.

Shift Supervisor: Also known as watch engineer, principal engineer; means the highest ranking representative of management always on duty at the nuclear facility. The shift supervisor exercises immediate management control over plant operating personnel. The shift supervisor reports to the plant operations manager or the plant manager.

Surveillance: Close observation of a person or group, especially of one under suspicion.

Suspicion: A condition in which one may reasonably suspect something ill, distrust, or doubt without proof or on slight evidence.

Tactical Response: That action taken by one or more guards when confronting one or more intruders inside the protected or vital area.

Vital Area:

Any area which contains vital equipment within a structure, the walls, roof, and floor of which constitute physical barriers of construction at least as substantial as walls as described in paragraph (f) (2) of this section 10CFR73.2 (h). See Figure 18.

Vital Equipment:

Any equipment, system device, or material, the failure, destruction, or release of which could directly or indirectly endanger the public health and safety by exposure to radiation. Equipment or systems which would be required to function to protect public health and safety following such failure, destruction or release are also considered to be vital. 10CFR73.2(i).

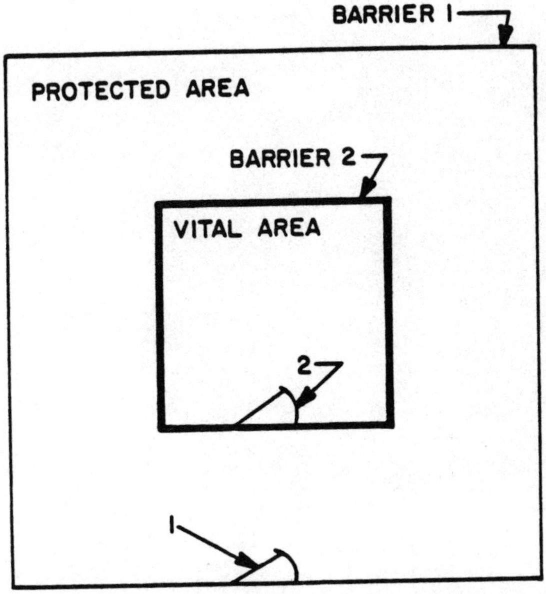

1. GATE TO, PROTECTED AREA
2. GATE OR DOOR TO VITAL AREA

Figure 29. Description of Protected Area and Vital Area